Radiocesium Dynamics in a Japanese Forest Ecosystem

Chisato Takenaka • Naoki Hijii
Nobuhiro Kaneko • Tatsuhiro Ohkubo
Editors

Radiocesium Dynamics in a Japanese Forest Ecosystem

Initial Stage of Contamination After the Incident at Fukushima Daiichi Nuclear Power Plant

Springer

Editors
Chisato Takenaka
Graduate School of Bioagricultural Sciences
Nagoya University
Nagoya, Aichi, Japan

Naoki Hijii
Graduate School of Bioagricultural Sciences
Nagoya University
Nagoya, Aichi, Japan

Nobuhiro Kaneko
Faculty of Food and Agricultural Sciences
Fukushima University
Fukushima, Japan

Tatsuhiro Ohkubo
School of Agriculture
Utsunomiya University
Utsunomiya, Tochigi, Japan

ISBN 978-981-13-8605-3 ISBN 978-981-13-8606-0 (eBook)
https://doi.org/10.1007/978-981-13-8606-0

This Springer imprint is published by the registered company Springer Nature Singapore Pte Ltd.
The registered company address is: 152 Beach Road, #21-01/04 Gateway East, Singapore 189721, Singapore

Preface

In March 2011, the East Japan was visited by the great strong earthquake. Not only a large number of earthquake victims but also all of the Japanese people were shocked by the news about the accident of the Fukushima Daiichi Nuclear Power Plant (FDNPP) caused by tsunami. From the uncontrollable nuclear reactors due to the loss of electricity by tsunami, a large amount of radioactive nuclides were emitted through the vent opening or the hydrogen explosion. The radionuclides dispersed to the surrounding area, especially to the northwest direction from the FDNPP, and deposited to the land surface. Since 70% of the Fukushima Prefecture is covered with forests, the contamination of forests with radionuclides has been a serious problem.

The forests in Fukushima include several types, such as deciduous broad-leaved forest, evergreen coniferous forest, and so on. The accident was occurred in March, when the leaves of deciduous trees have not developed. Therefore, in deciduous broad-leaved forests, most of the radionuclides directly deposited to the forest floor, and a part of radionuclide deposited to the trunks and branches of the trees. On the other hand, evergreen tree species, such as *Cryptomeria japonica*, had needle leaves in tree crown at the accident, and then more amounts of radionuclides deposited on the crown. This means that the amounts, translocation, and dynamics of radionuclide deposited in forests were different depending on the forest types constituted by the different tree species.

On the contamination of forests with radionuclides, many scientific papers related with the Chernobyl accident were published. However, the forest types, constituting trees, and forest ecosystems were different between the cases of Chernobyl and Fukushima. In the case of Fukushima, the researches on the radionuclides contamination of forests were started just a few months after the accident. The species of radionuclides and their chemical forms at the deposition were also different between the two accidents. Therefore, the researches on the Fukushima case should show new knowledges about the dynamics of radionuclides in a forest ecosystem.

Japanese forests near the residential area are called "Satoyama," where the specific culture has been developed. People enter the forests to get edible young tree leaves and bamboo shoot in spring season and mushrooms in autumn season. They used woods as fuel or bed log for mushroom. Also, the litters in forests have been used as fertilizer after composing. The radionuclide contamination affected these traditional food and life cultures and the organic agriculture using the composted litter. The radionuclide contamination restricts the activities by people in Satoyama. Therefore, the local people living near the contaminated Satoyama are very interested on when they will restart their cultural activity again.

The content of this book is a part of the outcome of the scientific research project entitled "Interdisciplinary Study on Environmental Transfer of Radionuclides from the Fukushima Daiichi NPP Accident" (Leader: Prof. Yuichi Onda) by the Grant-in Aid for Scientific Research on Innovative Areas sponsored by MEXT. This book is composed of four parts. In Part I, the radionuclide contamination of forests by the FDNPP accident is overviewed from the viewpoints of the comparison with the case of Chernobyl and the movements of radiocesium through hydrological processes. Part II focuses on the uptake process of radiocesium in plants through the analysis of field samples and the experiments using model plants. In Part III, the dynamics of radiocesium in a forest ecosystem, including not only plants but also microorganism, soil animals, spiders, and deer, are described. The future situation of the radiocesium contamination in forests and the relationship with the human activity, such as usage of forest products, are discussed in Part IV.

We must never cause any accident of nuclear power plant. But, unfortunately, if the similar situation will occur, the knowledge in this book will help to reduce the effects by radionuclides.

Nagoya, Aichi, Japan Chisato Takenaka

Contents

Part I
Radiocesium Deposition at the Accident

Chapter 1
Radioactive Contamination in Forest by the Accident of Fukushima Daiichi Nuclear Power Plant: Comparison with Chernobyl

Vasyl Yoschenko, Valery Kashparov, and Tatsuhiro Ohkubo

Abstract In this chapter, we compare the compositions and magnitudes of releases of radionuclides during the Chernobyl and Fukushima accidents and summarize the results of the long-term observations of the radionuclide dynamics in the ecosystem compartments in Chernobyl forests. Due to much larger magnitude of atmospheric release, the area contaminated as a result of the Chernobyl accident is larger; moreover, the near zone of the Chernobyl accident is contaminated with ^{90}Sr and other fuel component radionuclides that were not released in any significant amounts during the Fukushima accident.

Similarly to the forest ecosystems in Fukushima, the dominant process at the early stage after the deposition in Chernobyl forests was removal of the intercepted radionuclides from the aboveground forest biomass (foliage, bark) with litterfall and precipitations. In the following period, under certain conditions, the radionuclide concentrations and inventories in the aboveground biomass compartments started to increase till reaching the quasi-equilibrium levels in approx. 10 years after the accident. That was caused by an increase of the radionuclide bioavailable forms in the root-inhabited soil layer. At the late stage after the deposition, ^{137}Cs and ^{90}Sr in the Chernobyl forest ecosystems are involved into biological cycle: they are absorbed from soil, translocated and accumulated in the tree organs, and removed from the aboveground biomass by the same mechanisms that recycle their chemical analogs and essential plant nutrients K and Ca, respectively.

Keywords Chernobyl accident · Forest ecosystems · ^{137}Cs · Root uptake · Biological cycling

---

V. Yoschenko (✉)
Institute of Environmental Radioactivity, Fukushima University, Fukushima, Japan
e-mail: r705@ipc.fukushima-u.ac.jp

V. Kashparov
Ukrainian Institute of Agricultural Radiology, National University of Life and Environmental Sciences of Ukraine, Kyiv, Ukraine

T. Ohkubo
School of Agriculture, Utsunomiya University, Utsunomiya, Tochigi, Japan

C. Takenaka et al. (eds.), *Radiocesium Dynamics in a Japanese Forest Ecosystem*,
https://doi.org/10.1007/978-981-13-8606-0_1

1.1 Introduction

The accidents at the Fukushima Daiichi Nuclear Power Plant (FDNPP) on March 11, 2011, and the Chernobyl Nuclear Power Plant (ChNPP) on April 26, 1986, were the biggest radiation accidents in the humankind history. Both rated by the IAEA as of the INES Level 7 ("Major Accident"), the accidents resulted in releases of large amounts of radionuclides into the environment causing "widespread health and environmental effects" (IAEA 2013). The reasons of the accidents, compositions, and magnitudes of release and severity of their health and environmental consequences are different (Steinhauser et al. 2014). In context of this book, however, it is important that in both cases forests are the dominant component of the severe contaminated terrestrial environments. We believe that comparison of the results of the studies performed in the radioactive contaminated forests of Chernobyl and Fukushima can give better understanding of the mechanisms governing the radionuclide dynamics in the forest ecosystems.

Although the studies of the radionuclide dynamics in the forest ecosystems started in the early 1960s of the twentieth century (e.g., Auerbach and Ritchie 1963; Witkamp and Frank 1964; Waller and Olson 1967; Tikhomirov and Shcheglov 1994), the Chernobyl accident in 1986 became an event that caused fast development of forest radioecology. The increasing interest to forest radioecology in the post-Chernobyl period can be easily seen from the dynamics of the number of publications in this field that are listed in the Scopus database (https://www.scopus.com). For instance, we found 16 articles (including reviews) that were published in the period from 1969 to 1986 (during 18 years before Chernobyl) and included the words "forest" and "radionuclide" in the title, abstract, or keywords list (Fig. 1.1). In the next 18 years, till 2004, the number of such articles rose to 228, and that sharp increase was due to the articles containing words "Chernobyl" and "forest." Finally, by now (as of October 5, 2018), the total number of articles in the category "Chernobyl" and "forest" reached 413; among these articles, 348 were published in the fields of environmental, agricultural, and biological sciences.

The Fukushima accident stimulated the further development of forest radioecology. The number of articles including words "Fukushima" and "forest" reached 209 (with 170 in the fields of environmental, agricultural, and biological sciences) in 7 years after the accident (Fig. 1.1) vs 43 published during the same period in the category "Chernobyl" and "forest." However, many Chernobyl studies of the early period were reported in the Russian language journals only or were listed in the reviews published later in the international journals. In this chapter, we briefly summarize the main findings of the studies carried out in the radioactive contaminated Chernobyl forests which are relevant to content of this book (radionuclide deposition levels and dynamics in the forest ecosystems). Detail comparison of the impacts of the two accidents on forestry, economical loss, and effects of radiation on ecosystems can be found in our recent review (Yoschenko et al. 2018a).

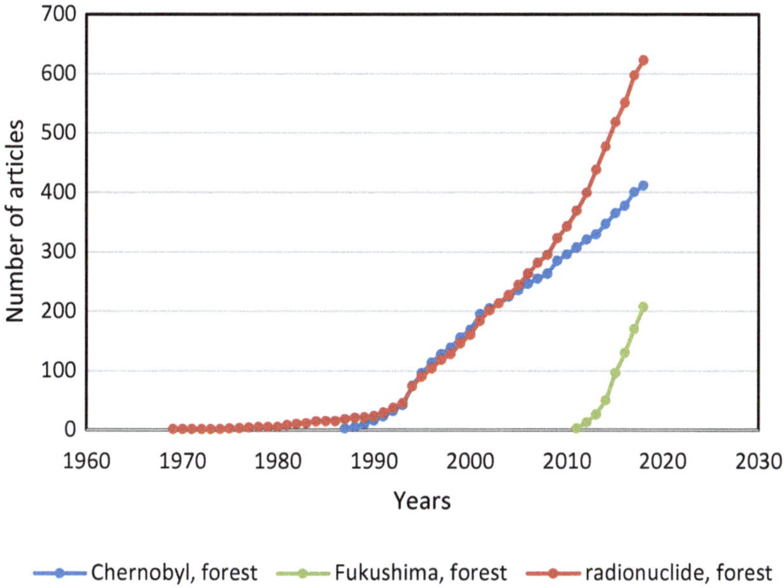

Fig. 1.1 Dynamics of the number of publications (articles and reviews) in the Scopus database illustrating the increases of the interest to forest radioecology after the two major radiation accidents

1.2 Atmospheric Releases and Terrestrial Depositions of Radionuclides: Fukushima vs Chernobyl

On April 26, 1986, two subsequent explosions ruined the reactor core and building of the ChNPP unit #4 and resulted in release of large amounts of volatile nuclear fission products (e.g., radionuclides of iodine, cesium) and noble gases and particles of irradiated nuclear fuel into the environment (Balonov 2007; Kashparov et al. 1996, 1999). The release continued during the fire in the destroyed reactor core that lasted for 10 days after the explosions (Izrael et al. 1990; Kashparov et al. 1999). Estimates of the radionuclide releases are shown in Table 1.1 (modified from Steinhauser et al. 2014). Here, we limit to the radionuclides that remain of radioecological importance in the Chernobyl exclusion zone or are interesting in context of comparison of the two accidents. Volatile long-lived radionuclides, such as ^{137}Cs, were transported in the atmosphere to the long distances and spread over the large territories in Russia, Belarus, Ukraine, Scandinavian countries, Austria, and, for the less extent, other European countries (Figs. 1.2 and 1.3). Radionuclides of the refractory elements were released in the fuel particles (FP) form (^{90}Sr, ^{95}Zr, ^{95}Nb, ^{99}Mo, ^{141,144}Ce, ^{154,155}Eu, ^{237,239}Np, $^{238–242}$Pu, ^{241,243}Am, ^{242,244}Cm). They deposited mainly in the Ukrainian part of the near zone of the accident (Figs. 1.4 and 1.5) forming several well-marked traces of fallout (Kashparov et al. 1999, 2001, 2003, 2018). The contamination maps (Figs. 1.3, 1.4, and 1.5) were created using the results of measurements of radionuclide activities in the soil

Table 1.1 Atmospheric releases of radionuclides from ChNPP and FDNPP (PBq)

Radionuclide	$T_{1/2}$[a]	Chernobyl	Fukushima
Noble gases and volatile elements			
^{85}Kr	10.76 years	33[b]	44[c]
^{133}Xe	5.24 days	6500[b]	14,000[d]
^{131}I	8.03 days	1760[e]	150[f]
^{137}Cs	30.08 years	85[e]	15–20[g]
Fuel component			
^{90}Sr	28.79 years	4.2[h]	0.02[i]
^{238}Pu	87.7 years	0.022[h]; 0.015[e]	2×10^{-6}–5×10^{-6}[i]
^{239}Pu	24,110 years	Total of 0.043[h]; 0.31[e]	
^{240}Pu	6561 years		
^{241}Pu	14.29 years	2.6[e]	1×10^{-6}–2.4×10^{-6}[j]

[a]https://wwwndc.jaea.go.jp/NuC/
[b]Dreicer et al. (1996)
[c]Ahlswede et al. (2013)
[d]Stohl et al. (2012)
[e]UNSCEAR (2008)
[f]Chino et al. (2011) (recommended by Steinhauser et al. (2014) as most cited estimate)
[g]Aoyama et al. (2016)
[h]Kashparov et al. (2003)
[i]Steinhauser et al. (2014)
[j]Zheng et al. (2012)

samples collected in 1997 in more than 1300 sampling points in the Ukrainian part of the Chernobyl exclusion zone (Kashparov et al. 2001, 2003, 2018). These data were used also for refinement of the estimates of releases of the fuel component radionuclides presented in Table 1.1. The measurement results and other relevant information (coordinates, land use, soil type and properties, etc.) are freely available from the online database https://catalogue.ceh.ac.uk/documents/782ec845-2135-4698-8881-b38823e533bf (Kashparov et al. 2018).

Deposition in the near zone of the large amounts of radionuclides of refractory elements (so-called fuel component of release) is the principal feature of the Chernobyl accident. Being encapsulated in the FP matrices (uranium oxides), these radionuclides initially were not available for migration in soil and uptake by plants. However, with weathering of FP under the environmental conditions, the radionuclides leached into soil solutions and became mobile and bioavailable. The rates of the FP dissolution depended on the composition and degree of oxidization of matrix (e.g., UO_2, UO_{2+x}, UZrO) and on the acidity of media (Kashparov et al. 1999, 2000, 2004; Kashparov 2002). We will demonstrate later that in context of the radioactive contamination of the forest ecosystems in the near zone of the Chernobyl accident, presence of ^{90}Sr in deposition is especially important.

Unlike the Chernobyl accident, there was no atmospheric release of the significant activities of radionuclides of refractory elements during the Fukushima accident (Table 1.1). According to the accident scenario described by Steinhauser et al. (2014), loss of power supply and consequent cooling pump failure due to the

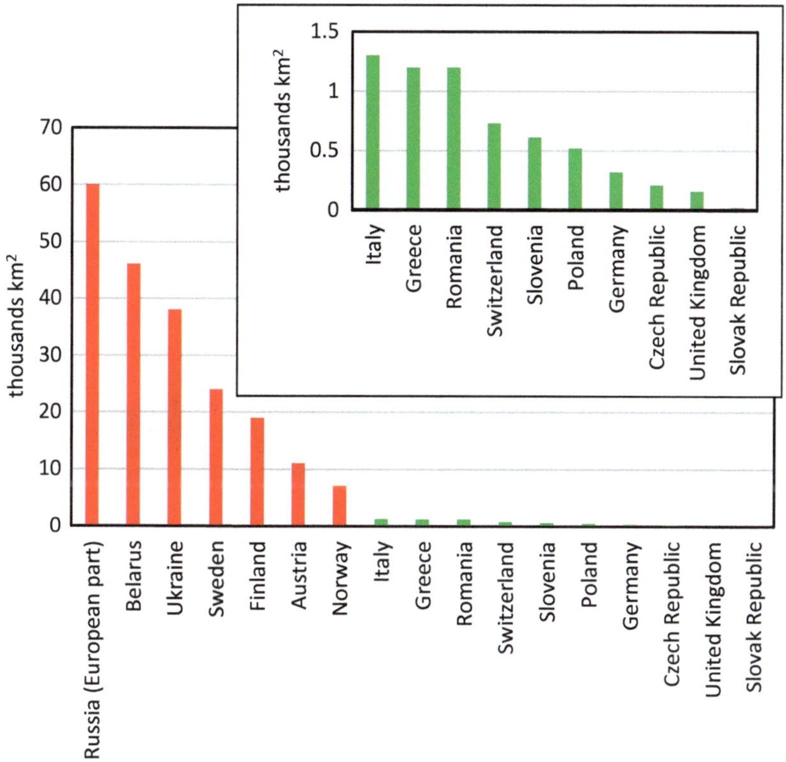

Fig. 1.2 Areas in Europe contaminated with ^{137}Cs above 40 kBq m^{-2}. (Data by De Cort et al. 1998)

tsunami on March 11, 2011, resulted in partial meltdown of nuclear fuel, increase of pressure, and formation of large amounts of hydrogen due to overheated vapor-zirconium reactions in the reactor cores of the units #1–4 of the FDNPP. Aimed to reduce pressure in the cores, venting procedures led to accumulation of hydrogen and radionuclides of noble gases and volatile elements in the reactor buildings followed with hydrogen explosions outside the cores and releases of the volatile elements radionuclides into the atmosphere. The temperatures in the reactor cores were not high enough for volatilization of the refractory elements, which explains the absence of radionuclides of these elements in the Fukushima fallout.

Atmospheric releases of the radionuclides of the volatile elements from the FDNPP were within the order of magnitude lower than from the ChNPP (e.g., 15–20 PBq of ^{137}Cs in Fukushima vs 85 PBq in Chernobyl; Table 1.1), and the major part of the Fukushima release was deposited to the ocean. Land deposition of ^{137}Cs, the only long-lived radionuclide released in significant amount from the FDNPP, is estimated as 3–6 PBq (Aoyama et al. 2016). Respectively, the contaminated area after the Chernobyl accident is much bigger than after the Fukushima accident. For example, the ^{137}Cs deposition exceeded 555 kBq m^2 at more than 10,000 km^2 after Chernobyl (Table 1.2) and exceeded 500 kBq m^{-2} at 646 km^2 in

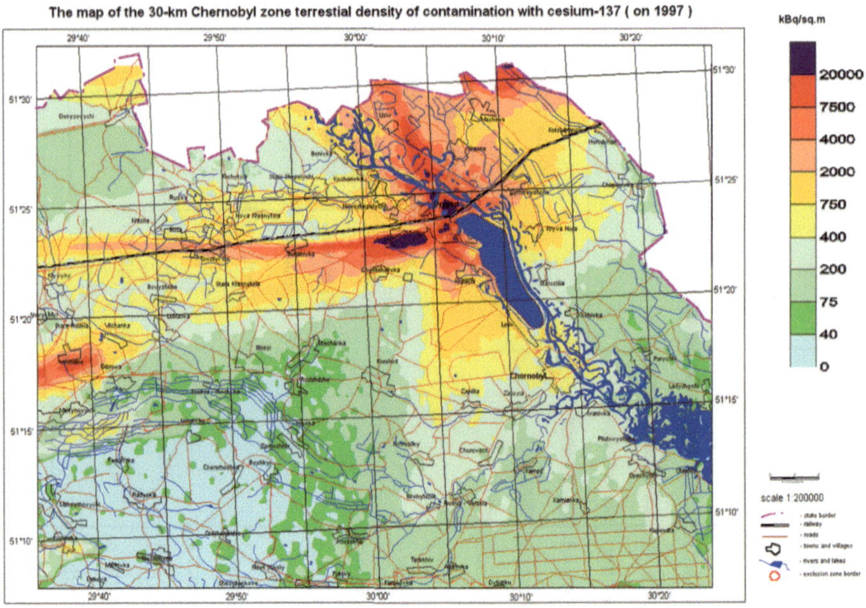

Fig. 1.3 Density of contamination with ^{137}Cs in the Chernobyl exclusion zone. (Ukrainian part; Kashparov et al. 2018)

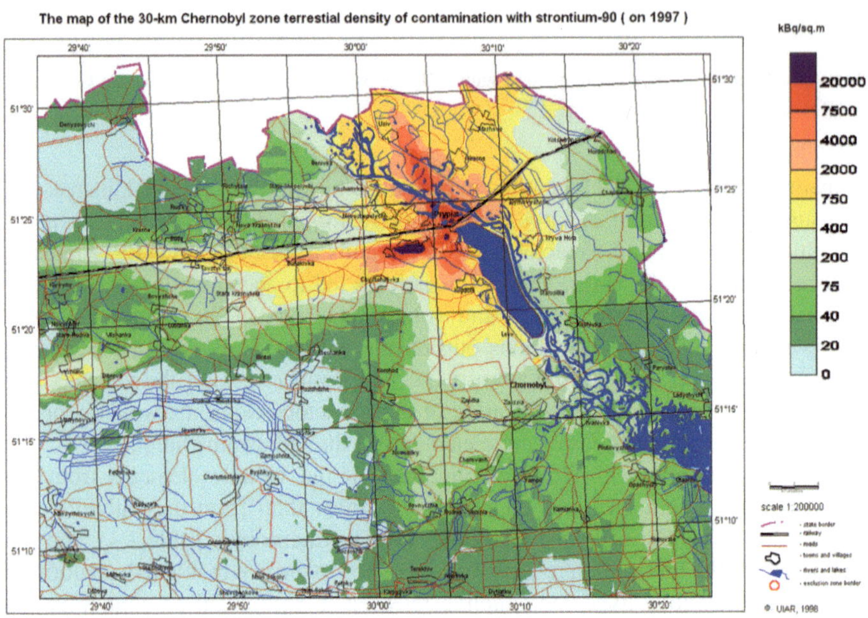

Fig. 1.4 Density of contamination with ^{90}Sr in the Chernobyl exclusion zone. (Ukrainian part; Kashparov et al. 2018)

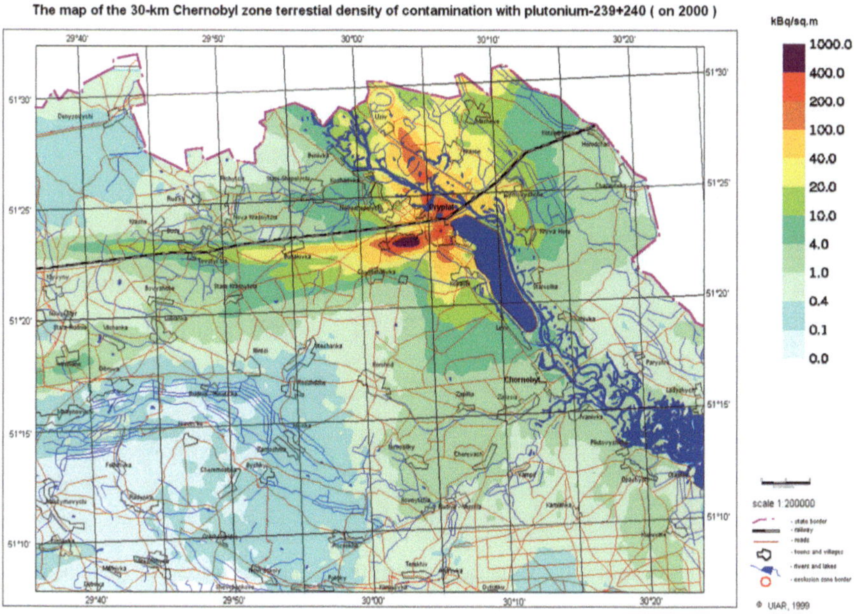

Fig. 1.5 Density of contamination with $^{239+240}$Pu in the Chernobyl exclusion zone. (Ukrainian part; Kashparov et al. 2018)

Table 1.2 The territory contamination with ^{137}Cs in Belarus, Russia, and Ukraine on October 05,1986 (in thousands km^2)

Country	Contamination density, kBq m^{-2}					Total	
	40–100	100–185	185–555	555–1480	>1480		
Russia	44	7.2	5.9	2.2	0.46	59.8[a]	31.1[b]
Belarus	21	8.7	9.4	4.4	2.6	46.1	
Ukraine	29	4.3	3.6	0.73	0.56	38.2[c]	21.5[d]

Estimated in 1998 (Ministry of Ukraine of Emergencies 2011; De Cort et al. 1998; Izrael and Bogdevich 2009)
[a]65,100 km^2 according to the estimation provided in 2006
[b]On 2006
[c]42,800 km^2 according to Ministry of Emergencies of Ukraine (2011)
[d]On 2011

Japan (Hashimoto et al. 2012); the deposition level of 185 kBq m^{-2} was exceeded at almost 30,000 km^2 after Chernobyl and at about 1700 km^{-2} after Fukushima (Ohta 2011). The ^{137}Cs deposition levels of 555 kBq m^{-2} and 185 km^{-2} are accepted in Ukraine as the criteria for classification of the territory as the zones of unconditional (mandatory) and guaranteed voluntary resettlement (Verkhovna Rada of Ukraine 1991).

Although there is such a big difference in the magnitudes of release and contaminated areas, ^{137}Cs deposition levels measured along the northwest trace of release (Fig. 1.6) are close to those in the near zone of the Chernobyl accident (Fig. 1.3).

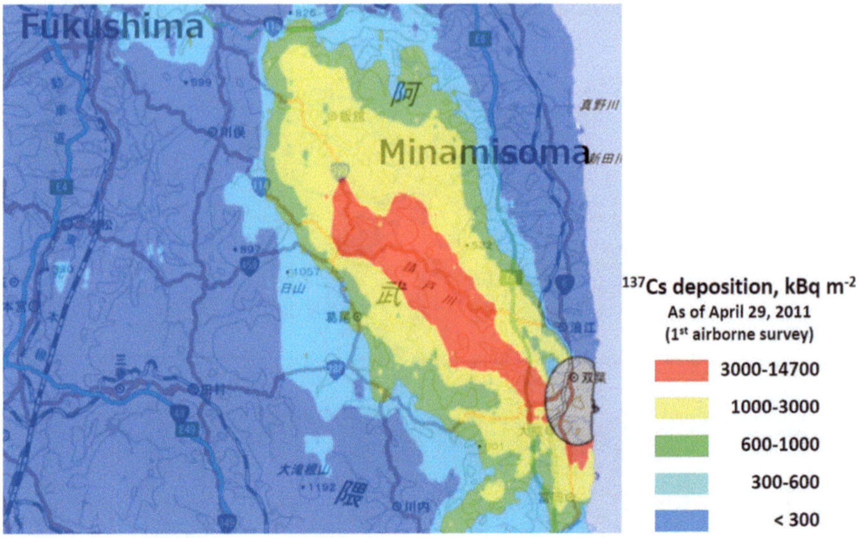

Fig. 1.6 Deposition densities of ^{137}Cs in the near zone of the Fukushima accident. (Modified from "Extension Site of Distribution Map of Radiation Dose, etc.,/GSI Maps". (http://ramap.jmc.or.jp/map/eng/). The map is quoted in accordance with the terms and conditions presented at http://ramap.jmc.or.jp/map/eng/about.html)

Unlike other radionuclides, the estimates of the released activities of ^{85}Kr and ^{133}Xe are larger for Fukushima than for Chernobyl (Steinhauser et al. 2014). It should be noted that according to Gusev and Belyaev (1991), ^{85}Kr contributes less than 1% into the total activity of the radioactive Kr isotopes present in the irradiated nuclear fuel of reactors RBMK-1000 (type of reactors at the ChNPP), while the rest is contributed by a series of isotopes with the half-life periods ranging from 32 s (^{90}Kr) to 4.48 h (^{85m}Kr). ^{133}Xe contributes about 35% to the total activity of Xe isotopes in the irradiated fuel, and about half of the total activity is formed by ^{138}Xe (14 min) and ^{139}Xe (41 s). Similarly, contribution of ^{131}I into the total activity of I isotopes is about 12%, while the major contributors are the short-lived isotopes with the half-life periods from 53 min (^{134}I) to 20.8 h (^{133}I). The actual contributions of short-lived radionuclides on the moment of explosion at the ChNPP could differ from those reported above because of the manipulations with the reactor power capacity during several hours preceding the accident. It is clear, however, that these radionuclides could be released in large amounts from the ruined reactor. According to Steinhauser et al. (2014), such short-lived radionuclides could greatly contribute to the environmental exposures formed in the first hours after the explosions in the vicinities of the ChNPP and did not contribute to such exposures in the case of the Fukushima accident because of their decay inside the reactors in the period between their emergency shutdown and releases. The accurate estimates of the dose rates formed by the short-lived radionuclides would require the very detailed information about the local meteorological conditions, height and temperature of release, etc. On the other hand, our simple calculations (Yoschenko et al. 2018a) show that even without accounting for the abovementioned short-lived radionuclides and α-emitting radionuclides, the dose rates near the FD1NPP in the initial period after the release should be significantly lower than in Chernobyl, which may explain the absence of effects of acute radiation to ecosystems in Fukushima (UNSCEAR 2015).

1.3 Forests in the Near Zones of the Accidents and Their Contamination Levels

According to the data by Davydchuk (1994), at the time of the accident, forests covered about 40% of the territory of the modern Chernobyl exclusion zone (Ukrainian art). The rest of the territory was presented by agricultural and meliorated lands (28% and 14%, respectively) and grassland and marsh areas (14%). The dominant tree species was Scots pine (*Pinus sylvestris* L.), which could be found at 80% of the afforested territory, while silver birch (*Betula pendula*) and common oak (*Quercus robur*) occupied 8–10% and 5–6% of the forest area, respectively (Davydchuk 1994). Scots pine was used for afforestation of this territory that began in the 1920s as a response to the sharp decrease of the land fertility (Ministry of Emergencies of Ukraine, National Academy of Sciences of Ukraine 1996; Nepyivoda 2005; Kuchma et al. 1997).

During three decades that passed since the Chernobyl accident, the forest area has increased to 58% of the total territory of the Chernobyl exclusion zone in Ukraine, reaching 1500 km^2 (Nikonchuk 2015) due to artificial afforestation and ecological succession of existing forest stands. Scots pine still is the dominant species, but now it covers only about 60% of the forest area, while the area occupied by birch has greatly increased to more than 25% (Table 1.3).

The forest territory distribution by the ^{137}Cs deposition levels to forests in three most affected countries after the Chernobyl accident is presented in Table 1.4. In total, ^{137}Cs deposition exceeded 555 kBq m^{-2} at the forest area of approx. 2630 km^2. For comparison, Hashimoto et al. (2012) estimated the forest total area in Japan with the ^{137}Cs deposition over 500 kBq m^{-2} as 428 km^2. Species distribution within this area is presented in Table 1.5. Unlike Chernobyl, deciduous broadleaf species occupy almost half of the total forest area; however, evergreen needleleaf species

Table 1.3 Current composition of forests in the Chernobyl exclusion zone (Nikonchuk 2015)

Species	Area, km^2	Wood volume, Mm^3
Conifers (mainly Scots pine)	892.5	22.2
Deciduous hardwood	81.5	1.5
Birch	385	4.2
Alder	100	1.8
Other	41	0.45
Total	1500	30
Ripe and overripe stands	86.3	2.2

Table 1.4 The forest territory contamination with ^{137}Cs in Belarus, Russia, and Ukraine (in thousands km^2) (Izrael and Bogdevich 2009; Nadtochy et al. 2003)

	Contamination density, kBq m^{-2}				
Country	37–185	185–555	555–1480	>1480	Total
Belarus[a]	12.9	3.2	1.7	0.14	17.9
Russia[b]	9.3	1.1	0.36	0.026	10.8
Ukraine[c]	10.9	1.06	0.31	0.095	12.3

[a]On 1.01.2006
[b]On 1.01.2006
[c]On 1.01.1993

Table 1.5 Tree species distribution in severe contaminated forests in Japan (Hashimoto et al. 2012)

Species	Area, km^2	Aboveground biomass, Mm^3
Deciduous broadleaf	210	4.3
Evergreen needleleaf	201	6.5
Deciduous needleleaf	17	0.5
Total	428	11.3

give the main contribution to the total aboveground biomass. The principal forestry species, Japanese cedar, is the dominant species in the evergreen needleleaf forests. The distributions of the forest areas by the levels of the ^{137}Cs deposition in each municipality of Fukushima Prefecture were recently reported by Kato and Onda (2018). Their results showed a huge variation of the deposition levels even at the municipality scale. In Namie Town and Futaba Town, the median ^{137}Cs depositions in forests reach 2252 kBq m^{-2} and 1459 kBq m^{-2}, i.e., are comparable to the deposition levels in the Chernobyl exclusion zone (>1480 kBq m^{-2} in Table 1.4). For the less extent, the ^{137}Cs deposition levels exceed 1480 kBq m^{-2} in forests in other municipalities (Iitate Village, Okuma Town, Tomioka Town, Minamisoma City, and Katsurao Village).

Thus, the ^{137}Cs deposition levels in forests in the near zones of the two accidents are close. However, in differ to Fukushima, in the Chernobyl zone, the areas with high levels of contamination with ^{137}Cs have been contaminated also with the fuel component radionuclides, particularly with ^{90}Sr (Figs. 1.3, 1.4, and 1.5). Its deposition levels range hundreds to thousands kBq m^{-2} at hundreds km^2 of forests in the Chernobyl exclusion zone (Table 1.6). Due to its comparable low absorption in the typical poor sandy soils of the Chernobyl zone, this radionuclide has high soil-to-plant transfer factors. At the close levels of deposition of ^{137}Cs and ^{90}Sr, the contributions of the latter one into contamination of biomass and doses to tree organs usually are much higher than those of ^{137}Cs (Yoschenko et al. 2006, 2011; Thiry et al. 2009). We will show further that the presence of ^{90}Sr in the radioactive deposition will remain an important factor limiting forestry in the near zone of the Chernobyl accident in the long term.

Table 1.6 Distribution of forest area in the Ukrainian part of the Chernobyl exclusion zone according to deposition of ^{90}Sr (as of 1997; based on the data from the online database (Kashparov et al. 2018))

Deposition, kBq m^{-2}	Area, km^2
<20	147
20–40	251
40–75	234
75–200	168
200–400	70
400–750	54
750–2000	57
2000–4000	21
4000–7500	8
7500–20,000	2
>20,000	0.005
Total	1012

1.4 Long-Term Radionuclide Dynamics in the Forest Ecosystems

In the recent review (Yoschenko et al. 2018a), we compared the main trends of dynamics of radionuclides in the ecosystem compartments of Chernobyl and Fukushima forests at the early stages after the deposition. The early-stage radionuclide dynamics in Fukushima forests along with the mechanisms of the radionuclide redistribution will be described in detail in the next chapters of this book, while in this chapter, we focus on the long-term radionuclide dynamics (including the early stage) in Chernobyl forests, present some example of current distributions of radionuclides in the forest ecosystems, and make a prediction of the future concentrations of radionuclides in wood in the Chernobyl exclusion zone.

The concept of stages of the radionuclide dynamics in the forest ecosystem compartments and description of the processes governing the radionuclide redistribution were summarized by Tikhomirov and Shcheglov (1994) based on the results of the long-term observations in the forest areas contaminated after the Kyshtym and Chernobyl accidents and by Shaw (2007).

At the first stage, the radionuclides that were released to the atmosphere have been deposited in different amounts onto the tree canopies and onto the forest floor. The initial partition of the deposited radionuclides between the aboveground biomass and forest floor depends on the forest type (species) and age, plantation density, season of vegetation period when deposition occurred, and type and physical-chemical forms of deposition (dry or wet, gaseous or particulate, etc.). For the Chernobyl release, the fraction intercepted by the tree canopies ranged from 40% of the total deposition in deciduous forests (Melin and Wallberg 1991) to 70–90% in conifer stands (Melin and Wallberg 1991; Bunzl et al. 1989; Tikhomirov and Shcheglov 1994).

The dominant process at the early stage after the deposition is removal of the intercepted radionuclides from the aboveground forest biomass. The radionuclides are leached out from foliage and bark with precipitations (via throughfall and stemflow) and moved from the canopies to the forest floor with litterfall (Shaw 2007). The radionuclide half-loss period from the tree canopies under various conditions ranged from 3–4 weeks to 4–6 months (Tikhomirov and Shcheglov 1994; Sombre et al. 1990). Bunzl et al. (1989) reported the two-exponential dependence for removal of Chernobyl-derived radiocesium from Norwegian spruce with the half-loss periods of 97 days (0.4 of the total deposition) and 195 days (0.26 of the total deposition).

The radionuclide removal processes continued at the later stages, too; however, the period of intensive removal of the initially intercepted radionuclide from the aboveground biomass, especially from the tree crowns, lasted for 2–4 years (note that longevity of leaves of the dominant species in Chernobyl forests, Scots pine, is 2–3 years). There are evidences that a part of initially intercepted radiocesium deposition may persist in the external bark for decades (Tsvetnova et al. 2018). However, in general, in 2–4 years the major fraction of the initially intercepted radionuclides was removed from the aboveground biomass and distributed between the forest litter and soil (Tikhomirov and Shcheglov 1994). Being distributed in soil, radionuclides became available for the root uptake by plants. The intensity of the root uptake of radionuclides into the aboveground plant biomass depends, among other factors, on the root distribution in soil (Fesenko et al. 2001). At that stage, the Chernobyl-derived radionuclides reached the root-inhabited soil layer. In certain conditions, it resulted in significant increase of the root uptake flux and, consequently, increase of the radionuclide activities in the aboveground biomass after the period of their decrease. The increase of the ^{90}Sr root uptake during the mentioned period was also related to leaching of the radionuclide from the fuel particles into bioavailable forms (Kashparov 2002). Such an increase of the radionuclide inventories in biomass and increase of soil-to-plant transfer factors to the tree compartments in the period between 1988 and 1992–1994 were reported in many studies, e.g., Shcheglov et al. (1996, 2001), Mamikhin et al. (1997), and Perevolotsky (2006). By the mid of the 1990s, the radionuclide concentrations in the forest compartments increased several times as compared to those observed at the end of the early stage. According to Tikhomirov and Shcheglov (1994), the quasi-equilibrium inventories in the biomass compartments were reached in 10–15 years after the deposition. For ^{137}Cs, the increase of its inventories in the aboveground biomass was well expressed mainly at hydromorphic soils and was less pronounced or absent in the forest growing in other conditions. Shcheglov et al. (2001) summarized the radiocesium dynamics types in Chernobyl forest ecosystems depending on the soil-landscape conditions:

- Increase of the transfer factors in hydromorphic soils due to the low fixation ability, intensive redistribution of radiocesium in the root-inhabited soil layer, and input of the radionuclide from the adjacent eluvial areas
- Decrease of the transfer factors in "automorphic soils of the eluvial landscapes" due to its removal from biomass and strong fixation to the soil particles
- Strong annual variability of the transfer factors without well-expressed general trends at semi-hydromorphic soils

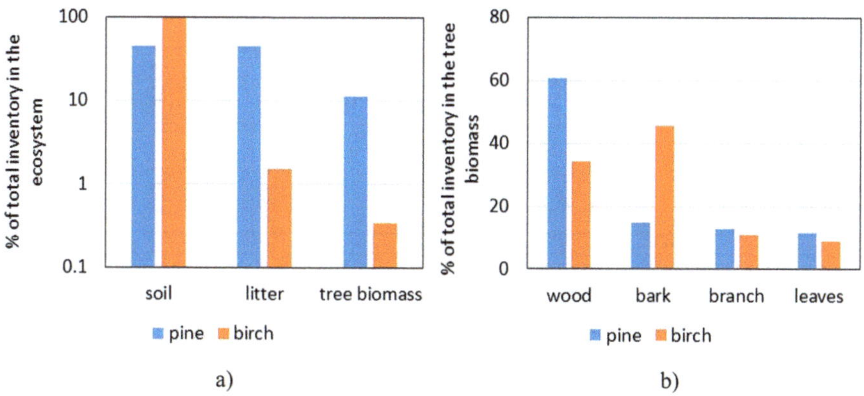

Fig. 1.7 ^{137}Cs distributions in the ecosystems (**a**) and in the tree biomass of pine and birch forests in the Chernobyl zone. (UIAR, unpublished data)

At the late stage after the deposition, involvement of ^{137}Cs and ^{90}Sr into biological cycle was the principal mechanism governing their dynamics in the Chernobyl forest ecosystems (Shcheglov et al. 2014). Essential role of the biogenic fluxes in formation of the radionuclide distributions in the forest ecosystems was demonstrated in our observations (unpublished data by Ukrainian Institute of Agricultural Radiology (UIAR)). We found that aboveground tree biomass and litter in pine forests in the Chernobyl zone currently can contain up to 10% and 40% of the total ^{137}Cs inventory in the ecosystem, respectively (Fig. 1.7). In differ, in deciduous forests, due to low annual increment of perennial biomass and fast rate of litter decomposition, more than 98% of the total ^{137}Cs inventory is localized in soil. The radiocesium soil-to-plant transfer factors (T_{ag}) in forests of the Chernobyl zone at the late stage after the deposition vary from 0.1 (Bq kg^{-1}) (kBq m^{-2})$^{-1}$ in dry fertile soils to 3 (Bq kg^{-1}) (kBq m^{-2})$^{-1}$ in wet low fertile soils (Krasnov et al. 2007).

As it was mentioned above, root uptake of ^{90}Sr into aboveground forest biomass depended on content of its bioavailable forms in soil that increased during the post-accidental period due to the radionuclide leaching from the fuel particles. Another important factor determining the magnitude of its root uptake is content of exchangeable calcium in soil. Calcium is essential element for plants and is absorbed from soil into biomass in the amounts necessary for the plant development. Being the chemical analog of Ca, strontium is absorbed from soil, translocated and accumulated in the tree organs, and removed from the aboveground biomass by the same mechanisms that recycle Ca. Since ^{90}Sr is present in soil in much less amounts than Ca, its soil-to-plant transfer factor decreases with increase of the exchangeable forms of Ca in soil (Fig. 1.8). In 30 years after the deposition, in forests growing at soils with low humus content, more than half of the whole ^{90}Sr inventory in the ecosystem can be localized in the aboveground tree biomass, while a large fraction, about 20%, has already migrated beneath the root-inhabited soil layer (data by UIAR).

Similarly to Sr, Cs is not required for plant development, but being a chemical analog of another essential plant nutrition element, potassium, Cs is absorbed from soil

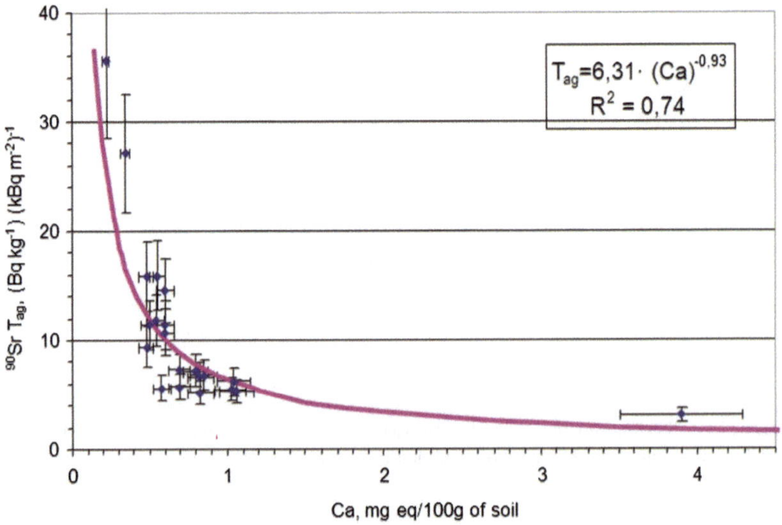

Fig. 1.8 Dependence of ^{90}Sr T_{ag} into pinewood on exchangeable calcium content in soil. (Data by authors (UIAR))

and redistributed within the forest biomass compartments by the mechanisms that recycle K in the ecosystem. Therefore, at the late stage after the deposition, the basic concepts used for modeling of the nutrients (Ca and K) cycling in forest ecosystems (Cole and Rapp 1981) may be applied for modeling of redistribution of radioisotopes of their chemical analogs, ^{90}Sr and ^{137}Cs, respectively. This approach was successfully applied for quantitative assessments of the annual radionuclide fluxes and for prediction of the long-term dynamics in Chernobyl forests at the late stage after the accident (Myttenaere et al. 1993; Goor and Thiry 2004; Goor et al. 2007; Thiry et al. 2009). In addition, in case of deposition of radiocesium on the forest ecosystems, important information for prediction of its long-term dynamics can be obtained from the distributions in the ecosystem of the natural stable Cs isotope, ^{133}Cs. At the late stage, when the residues of the initial deposition of ^{137}Cs are removed from the aboveground biomass and a quasi-equilibrium between its biogenic fluxes is reached, the distributions of radioactive and stable cesium isotopes should become similar, and their soil-to-plant transfer factors should be equal. This was demonstrated by Yoshida et al. (2002, 2004, 2011) for the range of the forest species studied in the radioactive contaminated ecosystems in Belarus, Europe, and Japan. Thus, knowledge about the current distributions of ^{133}Cs in Fukushima forests may shed a light on the future dynamics of the FDNPP-derived ^{137}Cs in the ecosystems (Yoschenko et al. 2018b).

Based on the above-described knowledge about the mechanisms and factors governing the radionuclide soil-to-plant transfer in the forest ecosystems at the late stage after the accident and taking into account the site-specific conditions (i.e., the radionuclide depositions, soil properties, etc.), UIAR made the predictions of compliance of the ^{90}Sr and ^{137}Cs concentrations in pinewood in the Chernobyl exclusion zone with the hygienic norms (published at http://uiar.org.ua/). The hygienic norms of ^{137}Cs in wood in Ukraine range from 600 Bq kg^{-1} in firewood to 3000 Bq kg^{-1}; for ^{90}Sr, the norm is 60 Bq kg^{-1} in firewood (Ministry of Health 2005). The observations and modeling results show that the ^{137}Cs concentrations in wood currently are below the strictest norm of 600 Bq kg^{-1} in the large part of the Chernobyl zone, and in 2100 wood from almost whole territory of the zone will comply the norm (Fig. 1.9). However, the ^{90}Sr concentrations in the present period almost everywhere exceed the hygienic norm and still will exceed it at the large part of the exclusion zone in the distant future (Fig. 1.10). Thus, presence of ^{90}Sr in the deposition in the near zone of the Chernobyl accident will limit the possibilities of economical utilization of forests in the long term.

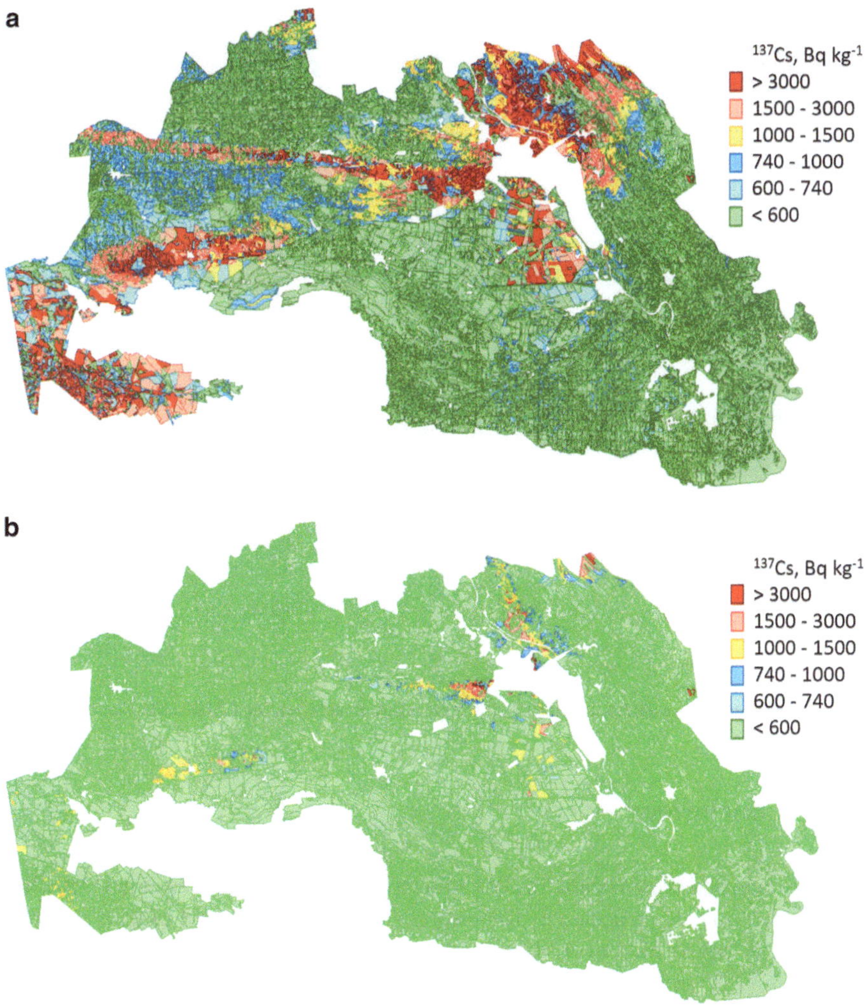

Fig. 1.9 Predicted concentrations of ^{137}Cs in pinewood in 2020 (**a**) and 2100 (**b**). (Data by authors (UIAR))

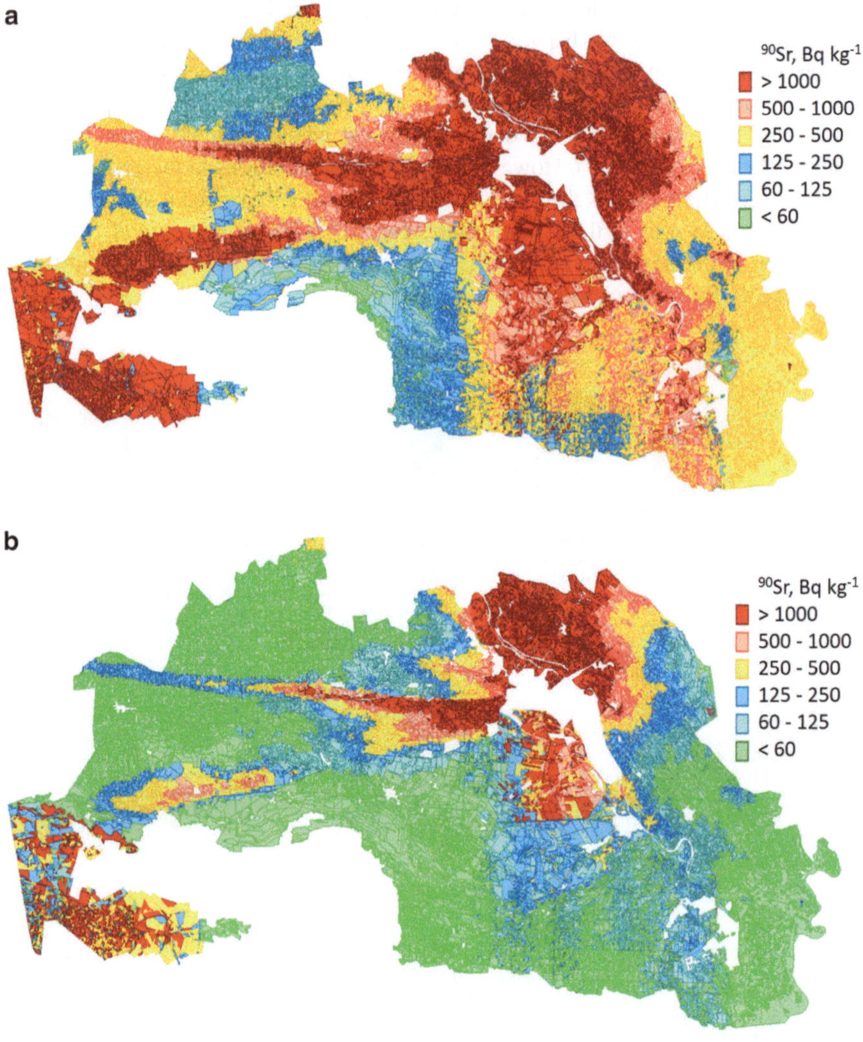

Fig. 1.10 Predicted concentrations of ^{90}Sr in pinewood in 2020 (**a**) and 2100 (**b**). (Data by authors (UIAR))

1.5 Conclusions

Forests cover the major parts of the radioactive contaminated areas after the Chernobyl and Fukushima accident. The area contaminated as a result of the Chernobyl accident is larger; moreover, the near zone of the Chernobyl accident is contaminated with ^{90}Sr and other fuel component radionuclides that were not released in any significant amounts during the Fukushima accident. Presence of ^{90}Sr in the deposition in the near zone of the Chernobyl accident will limit the possibilities of economical utilization of forests in the long term.

Similarly to the forest ecosystems in Fukushima, the dominant process at the early stage after the deposition in Chernobyl forests was removal of the intercepted radionuclides from the aboveground forest biomass (foliage, bark) with litterfall and precipitations. In the following period, under certain conditions, the radionuclide concentrations and inventories in the aboveground biomass compartments started to increase till reaching the quasi-equilibrium levels in 10–15 years after the accident (Tikhomirov and Shcheglov 1994). That was caused by increase of the radionuclide bioavailable forms in the root-inhabited soil layer. At the late stage after the deposition, ^{137}Cs and ^{90}Sr in the Chernobyl forest ecosystems are involved into biological cycle: they are absorbed from soil, translocated and accumulated in the tree organs, and removed from the aboveground biomass by the same mechanisms that recycle their chemical analogs and essential plant nutrients K and Ca, respectively.

Acknowledgment The authors' researches cited in this chapter were supported by the Japan Society for Promotion of Science: [Grant Numbers 15H00968, 15K00563, and 15H04621]; MEXT: [24110007]; and Science and Technology Center in Ukraine: [1992, 3674], the National University of Life and Environmental Sciences of Ukraine, under the Ministry of Education and Science of Ukraine [project 110/90-f]. We express our gratitude to Ms. Hiroko Nagata (IER) for editorial assistance.

References

Ahlswede J, Hebel S, Ross JO, Schoetter R, Kalinowski MB (2013) Update and improvement of the global krypton-85 emission inventory. J Environ Radioact 115:34–42

Aoyama M, Kajino M, Tanaka TY, Sekiyama TT, Tsumune D, Tsubono T, Hamajima Y, Inomata Y, Gamo T (2016) ^{134}Cs and ^{137}Cs in the North Pacific Ocean derived from the March 2011 TEPCO Fukushima Dai-ichi nuclear power plant accident, Japan. Part two: estimation of ^{134}Cs and ^{137}Cs inventories in the North Pacific Ocean. J Oceanogr 72:67–76

Auerbach SI, Ritchie JC (1963) Distribution of gamma-emitting nuclides in trees bordering radioactive waste pit No. 5. In: Health physics division annual progress report for period ending June 30, ORNL/3492. Oak Ridge National Laboratory, Oak Ridge

Balonov M (2007) The Chernobyl Forum: major findings and recommendations. J Environ Radioact 96:6–12

Bunzl K, Schimmack W, Kreutzer K, Schierl R (1989) Interception and retention of Chernobyl USSR-derived Cesium-134, Cesium-137 and Ruthenium-106 in a spruce stand. Sci Total Environ 78:77–78

Chino M, Nakayama H, Nagai H, Terada H, Katata G, Yamazawa H (2011) Preliminary estimation of release amounts of ^{131}I and ^{137}Cs accidentally discharged from the Fukushima Daiichi nuclear power plant into the atmosphere. J Nucl Sci Technol 48:1129–1134

Cole DW, Rapp M (1981) Elemental cycling in forest ecosystems. In: Reichle DE (ed) Dynamic properties of forest ecosystems. Cambridge University Press, Cambridge, UK, pp 341–407

Davydchuk V (1994) Landscapes of the accident zone and their radioactive contamination. In: Marinich A (ed) Landscapes of the Chernobyl zone and their evaluation regarding to conditions of the radionuclides migration. Naukova dumka, Kiev, pp 8–22

De Cort M, Dubois G, Fridman SD, Germenchuk MG, Izrael YA, Janssens A (1998) Atlas of cesium deposition on Europe after the Chernobyl accident, EUR Report Nr. 16733. Office for Official Publications of the European Communities; ECSC-EEC-EAEC, Brussels-Luxemburg

Dreicer M, Aarkrog A, Alexakhin R, Anspaugh L, Arkhipov NP, Johansson K-J (1996) Consequences of the Chernobyl accident for the natural and human environments. In: EC, IAEA, WHO (eds) One decade after Chernobyl: summing up the consequences of the accident. IAEA, Vienna, pp 319–361

Fesenko SV, Soukhova NV, Sanzharova NL, Avila R, Spiridonov SI, Klein D, Lucot E, Badot PM (2001) Identification of processes governing long-term accumulation of ^{137}Cs by forest trees following the Chernobyl accident. Radiat Environ Biophys 40:105–113

Goor F, Thiry Y (2004) Processes, dynamics and modelling of radiocaesium cycling in a chronosequence of Chernobyl-contaminated Scots pine (*Pinus sylvestris* L.) plantations. Sci Total Environ 325:163–180

Goor F, Thiry Y, Delvaux B (2007) Radiocaesium accumulation in stemwood: integrated approach at the scale of forest stands for contaminated Scots pine in Belarus. Environ Manag 85:129–136

Gusev N, Belyaev V (1991) Radioactive releases into the biosphere, Reference book, 2nd edn. Energoatomizdat, Moscow

Hashimoto S, Ugawa S, Nanko K, Shichi K (2012) The total amounts of radioactively contaminated materials in forests in Fukushima, Japan. Sci Rep 2:416

IAEA (2013) INES. The international nuclear and radiological event scale, User's manual 2008 edition. IAEA, Vienna

Izrael Y, Bogdevich I (eds) (2009) The Atlas of recent and predictable aspects of consequences of Chernobyl accident on polluted territories of Russia and Belarus (ARPA Russia-Belarus). Foundation "Infosphere" – NIA-Nature, Moscow

Izrael Y, Vakulovskiy S, Vetrov V, Petrov V, Rovinskiy F, Stukin E (1990) In: Izrael YA (ed) Chernobyl: radioactive contamination of natural environment. Book Company Hydrometeoizdat, Leningrad, p 296. (In Russian)

Kashparov V (2002) Assessment of the radiological situation resulted by the accidental release of fuel particles. Radioprotection Colloques 37:1061–1066

Kashparov VA, Ivanov YA, Zvarisch SI, Protsak VP, Khomutinin YV, Kurepin AD, Pazukhin EM (1996) Formation of hot particles during the Chernobyl nuclear power plant accident. Nucl Technol 114:246–253

Kashparov VA, Oughton DH, Zvarich SI, Protsak VP, Levchuk SE (1999) Kinetics of fuel particle weathering and ^{90}Sr mobility in the Chernobyl 30-km exclusion zone. Health Phys 76:251–259

Kashparov VA, Protsak VP, Ahamdach N, Stammose D, Peres JM, Yoschenko VI, Zvarych SI (2000) Dissolution kinetics of particles of irradiated Chernobyl nuclear fuel: influence of pH and oxidation state on the release of radionuclides in the contaminated soil of Chernobyl. J Nucl Mater 279:225–233

Kashparov VA, Lundin SM, Khomutinin YV, Kaminsky SP, Levchuk SE, Protsak VP, Kadygrib AM, Zvarych SI, Yoschenko VI, Tschiersch J (2001) Soil contamination with ^{90}Sr in the near zone of the Chernobyl accident. J Environ Radioact 56:285–298

Kashparov VA, Lundin SM, Zvarych SI, Yoschenko VI, Levchuk SE, Khomutinin YV, Maloshtan IM, Protsak VP (2003) Territory contamination with the radionuclides representing the fuel component of Chernobyl fallout. Sci Total Environ 317:105–119

Kashparov VA, Ahamdach N, Zvarich SI, Yoschenko VI, Maloshtan IM, Dewiere L (2004) Kinetics of dissolution of Chernobyl fuel particles in soil in natural conditions. J Environ Radioact 72:335–353

Kashparov V, Levchuk S, Zhurba M, Protsak V, Khomutinin Y, Beresford NA, Chaplow JS (2018) Spatial datasets of radionuclide contamination in the Ukrainian Chernobyl Exclusion Zone. Earth Syst Sci Data (ESSD) 10:339–353

Kato H, Onda Y (2018) Determining the initial Fukushima reactor accident-derived cesium-137 fallout in forested areas of municipalities in Fukushima Prefecture. J For Res 23(15):1–12

Krasnov V, Orlov A, Buzun V, Landin V, Shelest Z (2007) Applied forest radioecology. Polissiya, Zhytomyr

Kuchma M, Arkhipov A, Bidna S (1997) Specialized system of ecological-silvicultural maintenance of forests of the exclusion zone. Chernobyl Science and Technology Center of International Researches, Chernobyl

Mamikhin S, Tikhomirov F, Shcheglov A (1997) Dynamics of ^{137}Cs in the forests of the 30-km zone around the Chernobyl nuclear power plant. Sci Total Environ 193:169–177

Melin J, Wallberg L (1991) Distribution and retention of cesium in Swedish Boreal forest eco-systems. In: Moberg J (ed) The Chernobyl fallout in Sweden, results from a research programme on environmental radiology, The Swedish radiation protection project. Arprint, Lund, pp 467–475

Ministry of Emergencies of Ukraine (1996) National Academy of Sciences of Ukraine. Atlas of Chernobyl exclusion zone. Kartographiya, Kyiv

Ministry of Emergencies of Ukraine (2011) Twenty-five years after Chornobyl accident: safety for the future, National report of Ukraine. KIM, Kyiv

Ministry of Health of Ukraine (2005) Sanitary standard of specific activity of radionuclides ^{137}Cs and ^{90}Sr in wood and products of wood. Ministry of Health, Kyiv

Myttenaere C, Schell WR, Thiry Y, Sombre L, Ronneau C, van der Stegen de Schrieck J (1993) Modelling of the ^{137}Cs cycling in forest: recent developments and research needed. Sci Total Environ 136:77–91

Nadtochy P, Malynovsky A, Mozhar A, Lazarev M, Kashparov V, Melnyk A (2003) Experience of overcoming the consequences of the Chernobyl catastrophe (agriculture and forestry). Svit, Kyiv

Nepyivoda V (2005) Forestry in the Chornobyl exclusion zone: wrestling with an invisible rival. J For 103:36–40

Nikonchuk V (2015) Report for the round table "Exclusion zone: present and future" at SAUEZM on 13.11.2015. http://dazv.gov.ua/images/presentacii/Nikonchuk%20-%202015.11.13.pdf. Accessed 21 Dec 2017

Ohta H (2011) Environmental remediation of contaminated area by the Fukushima-Daiichi NPP accident. https://www.iaea.org/OurWork/ST/NE/NEFW/WTS-Networks/IDN/idnfiles/IDN_AnnFor2011/Cleanup_activities-OHTA.pdf. 2011

Perevolotsky A (2006) Distribution of ^{137}Cs and ^{90}Sr in the forest biogeocenoses. RNIUP, Gomel

Shaw G (2007) Radionuclides in forest ecosystems. Radioact Environ 10:127–155

Shcheglov A, Tikhomirov F, Tsvetnova O, Klyashtorin A, Mamikhin S (1996) Biogeochemistry of Chernobyl-derived radionucides in the forest ecosystems of the European part of the CIS. Radiation biology. Radioecol 36:469–478

Shcheglov A, Tsvetnova O, Klyashtorin A (2001) Biogeochemical migration of technogenic radionuclides in forest ecosystems. Nauka, Moscow

Shcheglov A, Tsvetnova O, Klyashtorin A (2014) Biogeochemical cycles of Chernobyl-born radionuclides in the contaminated forest ecosystems. Long-term dynamics of the migration processes. J Geochem Explor 144:260–266

Sombre L, Vanhouche M, Thiry Y, Ronneau C, Lambotte JM, Myttenaere C (1990) Transfer of radiocesium in forest ecosystems resulting from a nuclear accident. In: Desmet G (ed) Transfer of radionuclides in natural and seminatural environments. Elsevier, London, pp 74–83

Steinhauser G, Brandl A, Johnson TE (2014) Comparison of the Chernobyl and Fukushima nuclear accidents: a review of the environmental impacts. Sci Total Environ 470–471:800–817

Stohl A, Seibert P, Wotawa G (2012) The total release of xenon-133 from the Fukushima Dai-ichi nuclear power plant accident. J Environ Radioact 112:155–159

Thiry Y, Colle C, Yoschenko V, Levchuk S, Van Hees M, Hurtevent P, Kashparov V (2009) Impact of Scots pine (*Pinus sylvestris* L.) plantings on long term ^{137}Cs and ^{90}Sr recycling from a waste burial site in the Chernobyl Red Forest. J Environ Radioact 100:1062–1068

Tikhomirov F, Shcheglov A (1994) Main investigation results on the forest radioecology in the Kyshtym and Chernobyl accident zones. Sci Total Environ 157:45–57

Tsvetnova O, Shcheglov A, Klyashtorin A (2018) ^{137}Cs and K annual fluxes in a cropland and forest ecosystems twenty-four years after the Chernobyl accident. J Environ Radioact 195:79–89

UNSCEAR (2008) Sources and effects of ionizing radiation (annex D). United Nations, New York

UNSCEAR (2015) Developments since the 2013 UNSCEAR report on the levels and effects of radiation exposure due to the nuclear accident following the Great East-Japan earthquake and tsunami. A 2015 white paper to guide the Scientific Committee's future programme of work. United Nations, New York

Verkhovna Rada of Ukraine (1991) On the legal regime of the territories exposed to radioactive contamination in consequence of the catastrophe at the Chernobyl NPP. Bulletin of Verkhovna Rada, Kyiv, p 16

Waller HD, Olson JS (1967) Prompt transfers of cesium-137 to the soils of a tagged Liriodendron forest. Ecology 48:15–25

Witkamp M, Frank ML (1964) First year movement, distribution and availability of Cs in the forest floor under tagged tulip-poplars. Radiat Bot 4:485–495

Yoschenko V, Kashparov VA, Protsak VP, Lundin SM, Levchuk SE, Kadygrib AM, Zvarich SI, Khomutinin YV, Maloshtan IM, Lanshin VP (2006) Resuspension and redistribution of radionuclides during grassland and forest fires in the Chernobyl exclusion zone: part I. Fire experiments. J Environ Radioact 86(2):143–163

Yoschenko V, Kashparov VA, Melnychuk MD, Levchuk SE, Bondar YO, Lazarev M, Yoschenko MI, Farfan EB, Jannik GT (2011) Chronic irradiation of Scots pine trees (*Pinus sylvestris*) in the Chernobyl exclusion zone: dosimetry and radiobiological effects. Health Phys 101:393–408

Yoschenko V, Ohkubo T, Kashparov V (2018a) Radioactive contaminated forests in Fukushima and Chernobyl. J For Res 23(9):3–14

Yoschenko V, Takase T, Hinton TG, Nanba K, Onda Y, Konoplev A, Goto A, Yokoyama A, Keitoku K (2018b) Radioactive and stable cesium isotope distributions and dynamics in Japanese cedar forests. J Environ Radioact 186:34–44

Yoshida S, Muramatsu Y, Steiner M, Belli M, Pasquale A, Rafferty B, Rühm W, Rantavaara A, Linkov I, Dvornik A, Zhuchenko T (2002) Stable elements – as a key to predict radionuclide transport in the forest ecosystems. Radioprotection Colloques 37:C1-391–C1-396

Yoshida S, Muramatsu Y, Dvornik A, Zhuchenko T, Linkov I (2004) Equilibrium of radiocesium with stable cesium within the biological cycle of contaminated forest ecosystems. J Environ Radioact 75:301–313

Yoshida S, Watanabe M, Suzuki A (2011) Distribution of radiocesium and stable elements within a pine tree. Radiat Prot Dosim 146:326–329

Zheng J, Tagami K, Watanabe Y, Uchida S, Aono T, Ishii N, Yoshida S, Kubota Y, Fuma S, Ihara S (2012) Isotopic evidence of plutonium release into the environment from the Fukushima DNPP accident. Sci Rep 2:304

Chapter 2
Radiocesium Deposition at the Accident and the Succeeding Movement Through Hydrological Process in Forest Ecosystem in Fukushima

Hiroaki Kato

Abstract Fukushima Daiichi Nuclear Power Plant accident resulted in radioactive contamination of forest environment over a wide area in Fukushima Prefecture and the neighboring prefectures. In this chapter, initial atmospheric deposition of radiocesium following the Fukushima accident was estimated based on the analysis of multiple dataset derived from different airborne surveys. Furthermore, the canopy interception of fallout radiocesium by forest was reviewed and summarized according to the results reported in the previous studies and intensive field monitoring surveys by the author. The long-term dynamics of radiocesium in Japanese forests, such as transfer from the canopy to forest floor in association with hydrological and biological processes, were presented and discussed based on the field observation results by the author.

The forest area accumulated in total is 1.8 PBq of ^{137}Cs based on the analysis of the airborne monitoring surveys, which is corresponding to 72% of total ^{137}Cs activities deposited on the land area of Japan. The evergreen conifers tend to show high canopy interception rate greater than 70% of atmospheric input in most cases. This indicated that the canopy will act as a secondary source of radioactive contamination of the forest floor. On the other hand, the canopy interception by deciduous broad-leaved forest has not been sufficiently clarified because there have been limited data available for the canopy interception for deciduous broad-leaved species during growing season. The monitoring of radiocesium concentrations in hydrological and biological components effectively determined the transport of radiocesium from forest canopies; a double exponential field-loss model was used to simulate the observed loss of canopy radiocesium from Japanese cedar and konara oak forest mixed with red pine during the early phase of the accident. These results help to gain further understanding of key processes in transfer of atmospherically deposited radiocesium in forest ecosystems particularly during the early phase of the accident.

H. Kato (✉)
Center for Research in Isotopes and Environmental Dynamics, Laboratory of Advanced Research, University of Tsukuba, Tsukuba, Ibaraki, Japan
e-mail: kato.hiroaki.ka@u.tsukuba.ac.jp

C. Takenaka et al. (eds.), *Radiocesium Dynamics in a Japanese Forest Ecosystem*,
https://doi.org/10.1007/978-981-13-8606-0_2

Keywords Fukushima accident · Radiocesium · Forest · Canopy interception · Transfer

2.1 Radiocesium Fallout onto Forest Area After the Fukushima Daiichi Nuclear Power Plant Accident

The Fukushima Daiichi Nuclear Power Plant (FDNPP) accident resulted in the release of an enormous amount of radiocesium into the atmosphere (Chino et al. 2011; Amano et al. 2012; Hirose 2012), and the consequent atmospheric radiocesium fallout contaminated a large terrestrial area extending over Fukushima and its neighboring prefectures (Butler 2011; MEXT 2011a, b; NRA 2017). The contaminated area encompasses a wide range of environments and land uses (e.g., Kitamura et al. 2014), but approximately 70% of the total contaminated land area of Fukushima and its neighboring prefectures consists of forest (Hashimoto et al. 2012, 2013).

A series of airborne monitoring surveys of radioactivity have been conducted by the Japanese Ministry of Education, Culture, Sports, Science and Technology (MEXT) and provide basic information on radioactive contamination following the accident. Recent study reconstructed an initial fallout map of Fukushima accident-derived radiocesium based on a comparison of the deposition densities of the third and the fifth airborne monitoring surveys (Kato and Onda 2018; Kato et al. 2019). The deposition densities of the fifth survey were adjusted for variation in the measured radioactivity associated with the influence of multiple factors on radioactive decay, including the half-life of radiocesium, natural weathering processes, variation in the calibration procedure used between the airborne monitoring surveys, and other undefined mechanisms. Finally, calibrated deposition densities for each land use type from the fifth airborne monitoring survey were used to complement the fallout map derived from the third airborne monitoring survey (Fig. 2.1). The reconstructed fallout map covers all of Fukushima Prefecture and its neighboring prefectures. Furthermore, possible variation in radioactivity observations between the two airborne monitoring surveys was corrected based on a direct comparison of the deposition densities.

Total ^{137}Cs deposition in the eastern part of Japan was estimated as 2.5 PBq, based on a reconstructed fallout map of the Fukushima accident-derived ^{137}Cs (Table 2.1). This determination does not include ^{137}Cs deposition in the immediate area surrounding the Fukushima Daiichi Nuclear Power Plant; the total ^{137}Cs deposition onto land would increase to 2.7 PBq if ^{137}Cs deposition within 5 km of the reactor (0.2 PBq) was taken into account based on an unmanned helicopter measurement survey conducted by the JAEA. Forest areas accumulated 72% (1.8 PBq) of the total atmospheric input of ^{137}Cs to the land of Fukushima Prefecture based on the reconstructed fallout map (Fig. 2.2). The ^{137}Cs deposition density in forest areas showed significant variability among municipalities (Table 2.1).

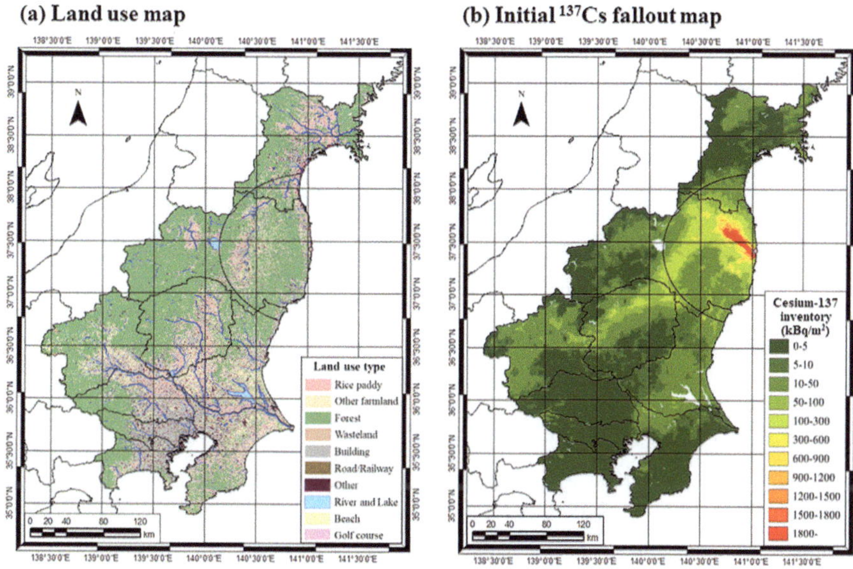

Fig. 2.1 Land use map (**a**) and the reconstructed initial ^{137}Cs fallout map following the Fukushima Daiichi Nuclear Power Plant accident (**b**) in eastern Japan (Kato et al. 2019)

Nevertheless, forest areas accumulated a large percentage of atmospherically deposited radiocesium in many municipalities. Statistical analysis of the variability in ^{137}Cs deposition density indicated that deposition density varied significantly, even within a municipality.

2.2 Radiocesium Cycling in Forest Ecosystems

Biogeochemical cycling of radiocesium in forest environment has been investigated following the past nuclear disasters such as the Chernobyl accident (Fig. 2.3). Each transfer process of radiocesium in and among different compartments has been well described in the IAEA documents (e.g., IAEA 2006).

Following atmospheric deposition of radiocesium onto forest area, the primary source of tree contamination was direct interception of aerosol-associated radiocesium by the canopy. Mechanical interception of radiocesium by canopy is followed by further translocation from foliar surfaces to structural components of the tree. Further changes in tree contamination after the initial fallout were due to two major processes. The first of these was a dominant and relatively rapid self-decontamination process of the tree canopy, in association with precipitation wash-off (throughfall) and litterfall. Once the canopy radiocesium reaches forest

Table 2.1 Deposition density of Fukushima-derived ^{137}Cs by land use type in each prefecture (Kato et al. 2019)

Prefecture	Paddy	Other farmland	Forest	Wasteland	Buildings	Roadway	Railway	Other	River lake	Beach	Golf	Prefecture total
Kanagawa	<1	<1	<1	<1	<1	<1	<1	<1	<1	<1	<1	<1
Tokyo	<1	<1	3	<1	1	<1	<1	<1	<1	<1	<1	4
Chiba	4	3	3	<1	8	<1	<1	2	1	<1	<1	21
Saitama	<1	<1	2	<1	1	<1	<1	<1	<1	–	<1	3
Gunma	2	7	83	1	3	<1	<1	1	1	–	1	99
Tochigi	20	9	114	2	9	<1	<1	2	3	–	2	161
Ibaraki	19	18	52	2	18	1	<1	3	3	<1	2	118
Fukushima	213	157	1473	10	92	3	3	15	22	<1	3	1992
Miyagi	11	5	58	2	4	<1	<1	1	2	<1	<1	84
Land use total	270	199	1789	17	136	5	4	24	32	<1	8	2482

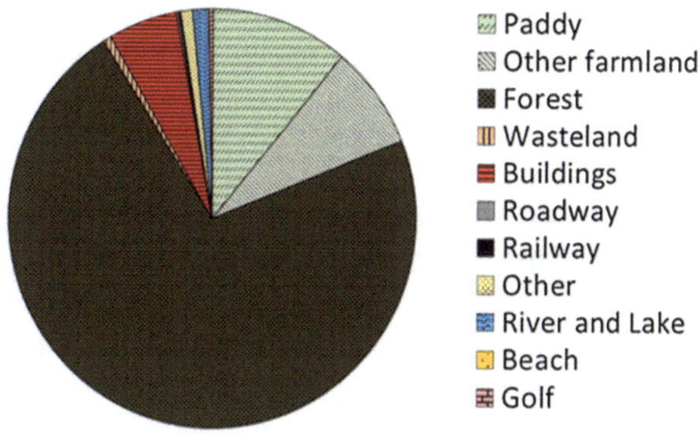

Fig. 2.2 The ^{137}Cs deposition (%) according to land use type (Kato et al. 2019)

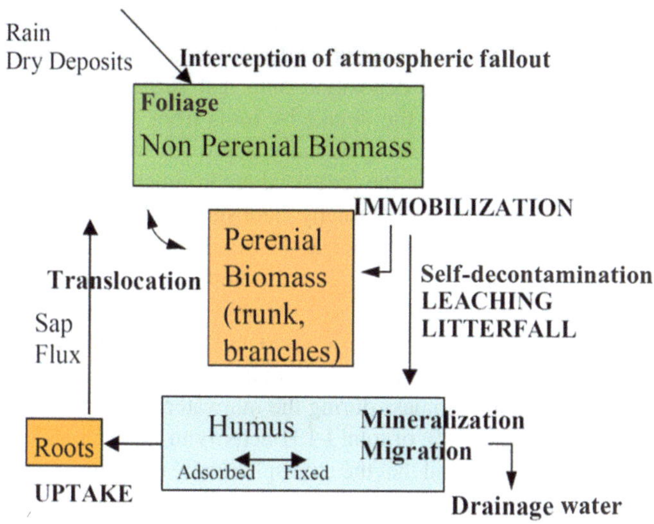

Fig. 2.3 Biogeochemical cycling of radiocesium in forest environment. (After IAEA 2006)

floor, it migrates into the soil profile over the longer term. This was followed by root uptake because radiocesium is a nutrient analogue to potassium. Therefore the rate of radiocesium cycling within forests is considered as relatively rapid, and quasi-equilibrium of its distribution is probably reached a few years after atmospheric fallout. Output from the system via the drainage water is generally limited as a result of radiocesium fixation on micaceous clay minerals. An important role of the vegetation in the recycling of radiocesium in the forest is the partial and transient storage of radiocesium, particularly in perennial woody components such as tree trunks and branches that can have a large biomass. The major portion of radiocesium accumulated by vegetation from the soil, however, is recycled annually through leaching and litterfall, resulting in the long-lasting biological availability of radiocesium in surface soil. Internal translocation of radiocesium within vegetation also occurs but involves generally low radiocesium activities compared with exchange (uptake/return) between the soil and the forest vegetation.

2.3 Initial Interception of Atmospherically Deposited Radiocesium by Forest Canopies

Forest canopy is an efficient filter of atmospheric contaminants due to huge surface area and aerodynamic roughness; therefore it acts as an important interface connecting the atmosphere and soil surface. Following a reactor accident, the primary source of tree contamination was direct interception of atmospheric radiocesium in association with both dry and wet deposition processes by forest canopies. Interception defines the fraction of dry or wet deposited radionuclides that is retained by vegetation. Since deposition onto leaf surfaces may cause a much higher plant contamination than the uptake by roots from contaminated soil, thus, initial canopy interception of atmospherically deposited radiocesium is probably most important to their transfer in the terrestrial environment and in food chains.

Canopy interception of atmospherically deposited radiocesium was observed following the Fukushima Daiichi Nuclear Power Plant accident. The rainwater volume and its radiocesium inventory were compared between open rainwater (RF) and throughfall (TF)/stemflow (SF) for the first rainfall event (during the period from March 11 to 28, 2011) following the accident (Fig. 2.4). The canopy interception of rainwater was 32% of total precipitation; however 93% of total ^{137}Cs activity was retained by the canopy biomass during the rainwater passing through the forest canopy. On the other hand, 50% of total I-131 activity in rainwater was captured by the canopy. These data indicated that the atmospherically deposited radiocesium was efficiently filtered and retained by forest canopies, whereas interception of I-131 was more or less smaller than radiocesium.

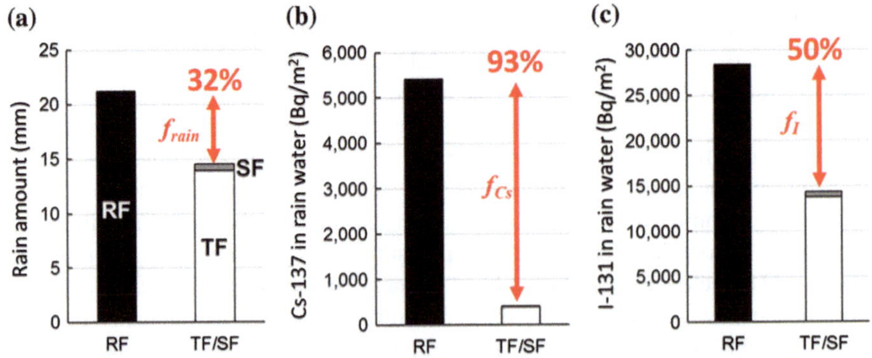

Fig. 2.4 Observed canopy interception of the Fukushima accident-derived radiocesium by evergreen Japanese cedar. (**a**) Rainwater, (**b**) Cs-137, (**c**) I-131

Interception factor *f* (dimensionless), which is defined as the ratio of the radionuclide inventory (Bq/m^2), is initially retained by the standing vegetation immediately subsequent to the deposition event, A_i, to the total deposition density (A_t, Bq/m^2).

$$f = \frac{A_i}{A_t} \tag{2.1}$$

The interception factor was observed and reported by several researchers for Japanese forests following the Fukushima reactor accident (Table 2.2). Although the methodology varied among studies (e.g., soil-litter sampling, mass balance, modeling, and experiment), the existing studies indicated that evergreen conifers tended to show higher interception factor for the atmospherically deposited radiocesium in contrast to the lower interception factor of deciduous broad-leaved species. The smaller number of sample collection (e.g., ≦3) likely resulted in huge variation of the determined interception factor in evergreen cedar forest.

Interception factor *f* for forest canopies has been reported for the forests affected by Chernobyl Nuclear Power Plant accident in 1986. The observed interception factor varied significantly among different tree species. Evergreen conifers (e.g., Norway spruce, red pine) tended to show higher interception factor than deciduous broad-leaved (e.g., beech, oak) forests (Table 2.3). This is because the initial fallout

Table 2.2 The observed interception factor for the Fukushima accident-derived radiocesium

Forest species	Radionuclides	f value (%)	Methodology	Reference
Japanese cedar	Cs-137, Cs-134	93	Mass balance (sample number 20)	Kato et al. (2012)
	I-131	51		
Japanese hinoki cypress	Cs-137, Cs-134	92	Mass balance (sample number 20)	Kato et al. (2012)
	I-131	25		
Japanese cedar	Cs-137, Cs-134	69	Soil and litter sampling (sample number 5)	Onda et al. (2015)
	I-131	29		
Japanese cedar	Cs-137	70	Field-loss model (sample number 7)	Kato et al. (2017)
Japanese hinoki cypress	Cs-137, Cs-134	44–45	Mass balance (sample number 3)	Itoh et al. (2015)
Japanese cedar	Cs-137, Cs-134	33–90	Mass balance (sample number 3)	Itoh et al. (2015)
Konara oak and red pine	Cs-137	23	Field loss (sample number 6)	Kato et al. (2017)
Deciduous broad-leaved	Cs-137, Cs-134	34	Mass balance (sample number 3)	Itoh et al. (2015)
Evergreen broad-leaved	Cs-137, Cs-134	34–44	Mass balance (sample number 3)	Itoh et al. (2015)

Table 2.3 The observed interception factor for the Chernobyl accident-derived radiocesium

Forest species	Radionuclides	f value (%)	Methodology	Reference
Coniferous	Not specified	70–90	–	Tikhomirov and Shcheglov (1991)
Coniferous	Cs-137	79	–	Ronneau et al. (1987)
Coniferous	Cs-137, Cs-134	70	Soil and litter sampling	Bunzl et al. (1989)
Norway spruce	Cs-137, Cs-134	70	–	Schimmack et al. (1991)
Coniferous	Cs-137	80–100	–	Melin et al. (1994)
Coniferous	Cs-137	80	–	Sombre et al. (1990)
Norway spruce	Cs-134	79–86	Experiment	Thiry et al. (1997)
Norway spruce	Cs-134	34	Mass balance (experiment)	Thiry et al. (2016)
Deciduous	Cs-137	10–40	–	Melin et al. (1994)
Beech	Cs-137, Cs-134	20	–	Schimmack et al. (1991)

occurred in the early spring, during the leafless season of many deciduous species for the Chernobyl accident. The situation was similar for the Fukushima accident where the initial atmospheric fallout occurred during the late winter to early spring. Therefore, it should be noted that the observed f for deciduous forest species represented initial canopy interception during the leaf-off season of the year; canopy interception by deciduous broad-leaved forest has not been sufficiently clarified because there have been limited data available for the canopy interception for deciduous broad-leaved species during the growing season.

Aggregate transfer factor (T_{ag}: m^2/kg), which is defined as Eq. 2.2, has been used as an indicator of radiocesium transfer from the soil to vegetation assuming the quasi-equilibrium condition of radiocesium cycling in forest ecosystems:

$$T_{\mathrm{ag}} = \frac{C_{\mathrm{veg}}}{A_t} \tag{2.2}$$

where C_{veg} is the activity of radiocesium in plants and A_t is the initial aerial deposition density. However, aggregate transfer factor can be used as an indicator representing the efficiency of canopy interception during the early period of the accidents when influence of direct contamination is dominant.

The Ministry of Agriculture, Forestry and Fisheries of Japan has conducted regional measurements of radioactive contamination in various forest environments in Fukushima Prefecture during August–October, 2011 (MAFF 2011). The results of regional survey outputted dataset of radiocesium inventories in litter and soil at 395 locations. On the other hand, Fukushima Prefecture measured radiocesium concentrations in different parts of the Japanese cedar tree at 85 locations in 2011 (Fukushima Prefecture 2016). Those data allow us to analyze initial canopy

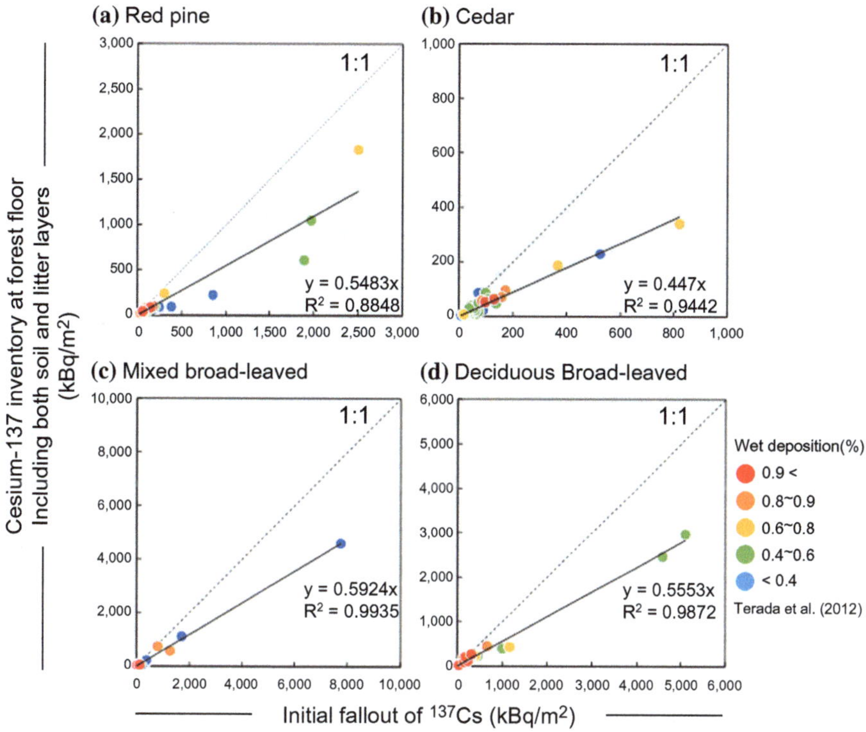

Fig. 2.5 Radiocesium inventory at forest floor of different tree species. (**a**) Red pine, (**b**) cedar, (**c**) mixed broad-leaved, (**d**) deciduous broad-leaved

interception of Fukushima accident-derived radiocesium by various forest species and its spatial dependency within Fukushima Prefecture. The ^{137}Cs inventory at the forest floor was plotted against initial atmospheric deposition following the Fukushima accident (Fig. 2.5). The slope of regression line indicates differences of initial canopy interception of ^{137}Cs by different tree species. The evergreen Japanese cedar shows lowest ^{137}Cs inventory at the forest floor, suggesting the efficient canopy interception of atmospherically deposited radiocesium up to 10–15% than that by deciduous broad-leaved forests.

Aggregate transfer factor (T_{ag}) for Japanese cedar needle was plotted over the wet deposition map derived from the model calculation of WSPEEDI by the JAEA (Katata et al. 2015) (Fig. 2.6). Although there are uncertainties in determination of initial fallout of ^{137}Cs at each sampling site, the calculated T_{ag} showed variations in two orders of magnitude at site by site. These regional patterns of T_{ag} suggested that initial direct contamination of tree parts varied site by site even within Fukushima Prefecture. Further investigation is required to clarify the causes (e.g., stand properties, deposition type, chemical form of fallout, etc.) producing such heterogeneity in radioactive contamination of forest components.

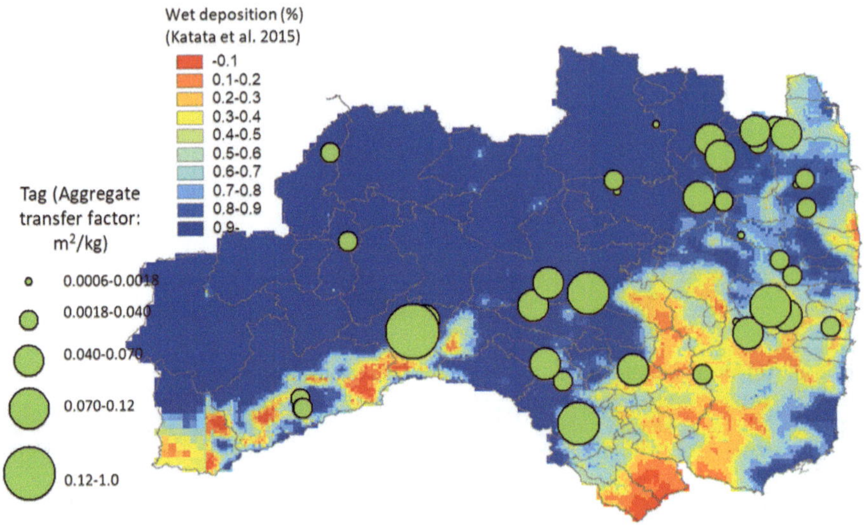

Fig. 2.6 Aggregate transfer factor of radiocesium in Japanese cedar needles

2.4 Radiocesium Transport from Canopy to Forest Floor

Initial canopy interception of atmospherically deposited radiocesium is followed by removal of contamination due to self-decontamination processes such as rainwash and mechanical breakdown of plant bodies as litterfall. Radiocesium transfer from the canopy to forest floor has been intensively monitored by the authors since the very early period of the Fukushima accident. The study sites are located 40 km northwest of the FDNPP and are highly contaminated by deposited radionuclides (Fig. 2.7). The plume released by the FDNPP from approximately 12:00 to 15:00 JST (Japan Standard Time) on March 15 flowed northwestward, and wet deposition via precipitation occurred on the same night (Chino et al. 2011). In the area, the total atmospheric deposition of ^{137}Cs following the FDNPP accident was estimated as 300–600 kBq m^{-2} based on results of the Third Airborne Monitoring Survey of radioactive contamination (MEXT 2011a, b).

Three forest stands were selected as experimental sites, namely, a mature (31 years old) cedar (*Cryptomeria japonica*) stand, young (15 years old) cedar stand, and mixed broad-leaved stand (Table 2.4). The distance between the cedar and mixed broad-leaved stands was approximately 2 km. By July 2, 2011, the total ^{137}Cs deposition after the reactor accident was estimated as 442 kBq m^{-2} in the mature and young cedar stands and 451 kBq m^{-2} in the broad-leaved stand based on the Third Airborne Monitoring Survey of radioactivity (MEXT 2011a, b).

Foliage (needles/leaves) and outer bark were sampled from the cedar stands and mixed broad-leaved forest. The foliage samples were collected from trees at different heights (5–15 m) by using a monitoring tower established in each experimental forest. Conversely, new leaves of konara oak tree were collected from various

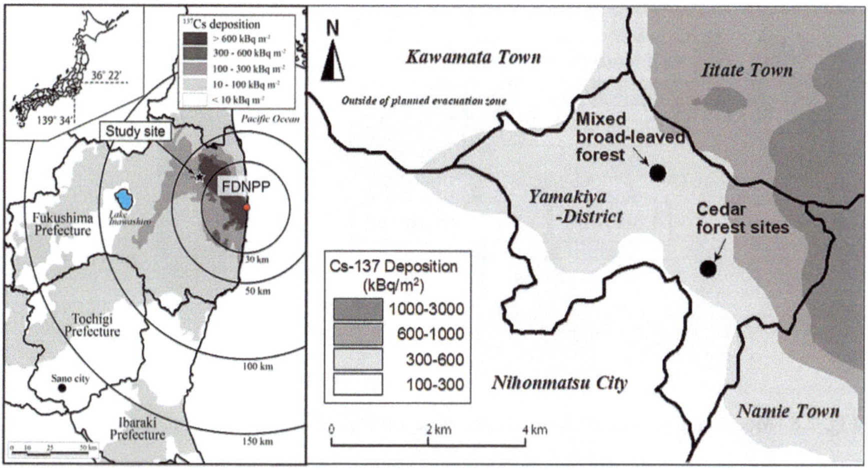

Fig. 2.7 Aggregate transfer factor of radiocesium in Japanese cedar needles

Table 2.4 Properties of experimental forest sites

	Mature cedar (MC)	Young cedar (YC)	Mixed broad-leaved (BL)
Stand age (year)	31	18	–
Stem density (stem/ha)	1250	2600	2500
Tree height (m)	25<	<20	25<
Slope (degree)	20	33	15
LAI (m^2/m^2)	4.2	10.3	–
^{137}Cs deposition (kBq/m^2)	442		451

heights. It is noted that needles and leaves were collected from at least three independent trees. The arithmetic mean of the measured ^{137}Cs activities at different heights was used as the representative value for the experimental forest. It should be noted that konara oak trees were selected as the target tree in the mixed broad-leaved forest because the konara oak trees are a dominant tree species and covered a large area of the experimental plot (>75%). In the cedar stands, new needles (which were defined as foliage developed during each sampling year) were separately collected. The remaining needles were considered as old foliage that developed before the sampling year. Conversely, in the broad-leaved forest, all the collected foliage was considered as newly developed leaves in each sampling year. The outer bark at breast height was collected from at least three different trees in the experimental plot, and bark surfaces in different directions were selected for sampling to minimize the influence of heterogeneous initial contamination.

Hydrological variables including open rainfall, throughfall, and stemflow were collected in the three forest stands at intervals ranging from 2 weeks to 2 months from July 2011 to March 2017 (Fig. 2.8). Seven throughfall collectors were placed in

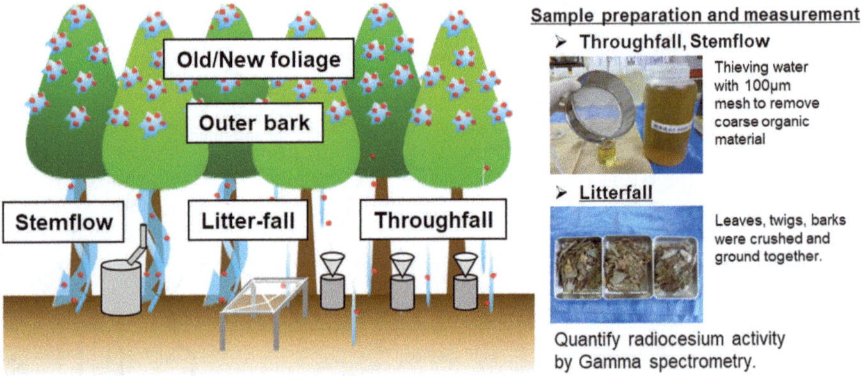

Fig. 2.8 Monitoring of radiocesium transfer via hydrological and biological pathways

a lattice-like pattern in each experimental plot of the young and mature cedar stands. Six throughfall collectors were randomly placed across the experimental plot in the mixed broad-leaved stand. Additionally, litterfall was collected at three locations in each experimental plot at intervals of several weeks to months by using a 1 m^2 PVC pipe frame. The details of the sampling were reported in Kato et al. (2017).

The old cedar needles and the outer bark of both the cedar and Japanese konara oak showed markedly high ^{137}Cs concentrations, indicating contamination by direct deposition of atmospheric ^{137}Cs onto the plant surfaces (canopy interception) following the Fukushima accident (Fig. 2.9a). The new needles/leaves of the cedar and Japanese konara oak showed high ^{137}Cs concentrations, although these were not affected by direct deposition from initial atmospheric fallout (Fig. 2.9b). The existence of ^{137}Cs in the new needles collected during 2011 indicated its rapid translocation within the cedar trees because of direct incorporation of ^{137}Cs from the foliar and bark surfaces. Foliar uptake had only a minor influence on ^{137}Cs concentrations in the new leaves of the Japanese konara oak, because these leaves had not developed at the time of the initial atmospheric fallout. Therefore, the ^{137}Cs uptake into the newly developed Japanese konara oak leaves possibly occurred through the outer bark surfaces of the stems, branches, and twigs, rather than through foliar uptake.

Cesium-137 concentrations in the cedar needles showed an exponential decrease, indicating self-decontamination of ^{137}Cs over time (Fig. 2.9). In contrast to the cedar needles, ^{137}Cs concentrations in the Japanese konara oak leaves were more or less constant over time. Cesium-137 concentrations in the outer bark samples (Fig. 2.10) showed a trend opposite to that of the needle/leaf concentrations, with more or less constant ^{137}Cs concentrations in the cedar and a significantly decreasing trend in the Japanese konara oak. The declining trend of the measured ^{137}Cs concentrations in foliage and outer bark samples was approximated by using a single exponential equation (Eq. 2.3) to calculate environmental half-life of ^{137}Cs:

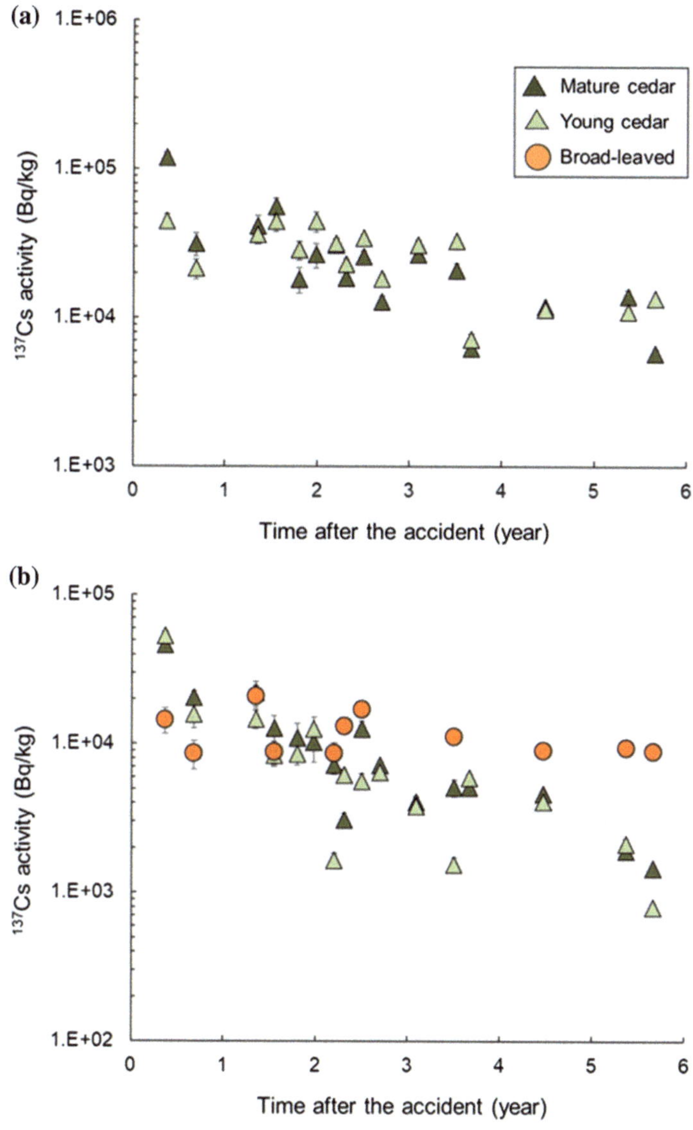

Fig. 2.9 Temporal change of radiocesium concentration in foliage samples. (**a**) Old needle, (**b**) new foliage

$$A(t) = A_0 e^{-\lambda t} \tag{2.3}$$

where $A(t)$ and A_0 are the ^{137}Cs concentration at time t and $t = 0$ and λ is the coefficient of reduction in ^{137}Cs concentration. The environmental half-life ($T_{1/2}$) can be calculated by using the following equation (Eq. 2.4):

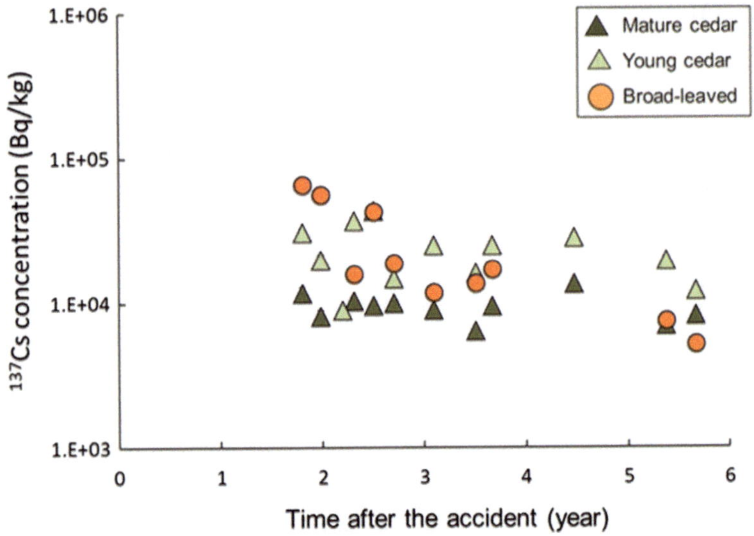

Fig. 2.10 Temporal change of radiocesium concentration in outer bark samples

$$T_{1/2} = \ln 2/\lambda \tag{2.4}$$

The environmental half-life for the Japanese cedar was 2.0–3.0 years for new needle, 2.2–2.4 years for old needle, and greater than 6.9 years for the outer bark. On the other hand, the environmental half-life for the konara oak was 6.3 years for new foliage and 1.5 years for outer bark. A previous study reported that the environmental half-lives of cedar needles and konara oak leaves were 0.7–1.1 years and 7.4 years, respectively (Imamura et al. 2017). On the other hand, those values for bark sample were less than 3.5 years for the Japanese cedar and 3.3–8.7 years for the konara oak tree (Imamura et al. 2017). The estimated environmental half-lives for foliage samples were comparable with the results of the previous study for konara oak leaves but slightly greater than the values previously reported. The stand age of their studied cedar forest was in the range of 38–57 years old; it is speculated that a difference in stand age affected the environmental half-life of the ^{137}Cs concentration in the cedar needles (e.g., Rauret et al. 1994). On the other hand, the environmental half-life of ^{137}Cs concentrations measured in the outer bark was greater for the cedar bark but smaller for the konara oak tree. Nevertheless, the environmental half-life of ^{137}Cs concentration in the outer bark of konara oak tree showed large variations.

The ^{137}Cs concentration in litterfall exponentially decreased over the 6-year observation period (Fig. 2.11a). The environmental half-life for litterfall was in a range of 2.1–2.2 years for Japanese cedar and 1.5 years for the mixed broad-leaved forest. These constants for the litterfall were comparable to the arithmetic means of those for the old and new needles of Japanese cedar. However, in the mixed broad-leaved forest, a distinct decreasing trend was in contrast to the constant

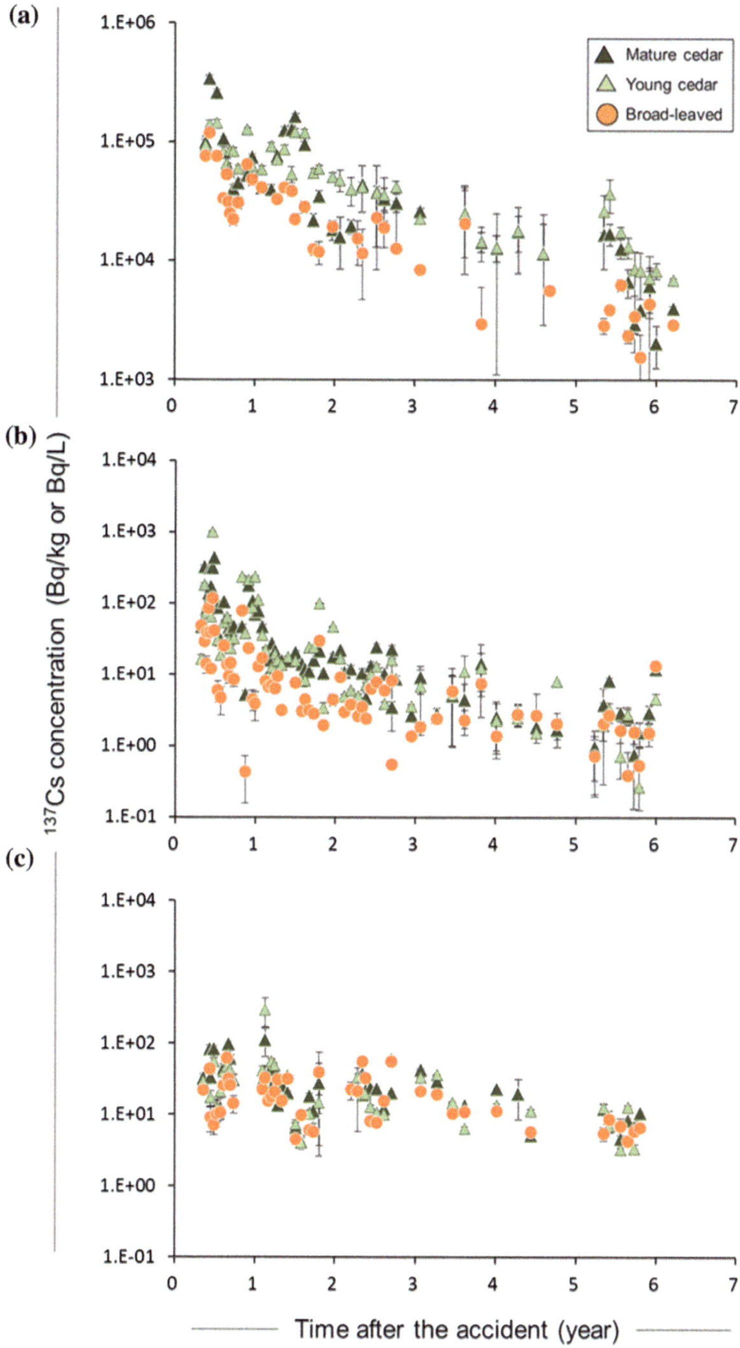

Fig. 2.11 Temporal change of radiocesium concentration in litterfall, throughfall, and stemflow. (**a**) Litterfall, (**b**) throughfall, (**c**) stemflow

concentrations found in the new leaves of the Japanese konara oak. The litterfall collected in the mixed broad-leaved forest was in fact a mixture of Japanese konara oak leaves, red pine needles, and the twigs and bark of these trees. Hisadome et al. (2013) reported the data showing the contribution of konara oak and red pine litterfalls to the observed litterfall mass (kg m^{-2}) and ^{137}Cs deposition (kBq m^{-2}) in the same mixed broad-leaved forest for an observation period from July 31, 2011, to May 25, 2012. The litterfall mass originating from the konara oak tree accounted for 74% of the total litterfall; however, contributions of these two tree species to the ^{137}Cs depositional flux via litterfall were nearly equal. It is suggested that the decreasing constants obtained in this study were affected by the rapid decreasing trend of the red pine. In addition, a marked seasonal variability in ^{137}Cs concentrations in Japanese konara oak leaves has been reported in previous studies (e.g., Hisadome et al. 2013; Okada et al. 2015). The ^{137}Cs concentration in Japanese konara oak leaves peaked during the growing season between May and September; however, the concentration significantly decreased during autumn. It is speculated that resorption of ^{137}Cs from senescing leaves may have occurred before leaf fall during autumn because of the leaf-fall phenology and leaf chemistry, as has been reported for potassium (e.g., Niinemets and Tamm 2005). Nevertheless, ^{137}Cs concentrations in the newly developed Japanese konara leaves have remained high for the last 6 years even though the canopy ^{137}Cs inventory has been reported to have decreased with time because of self-decontamination processes based on the direct measurement (Imamura et al. 2017) and mass balance of ^{137}Cs in the forest canopy (Kato et al. 2017). It is speculated that this discrepancy indicates the contribution of root uptake of ^{137}Cs from the soil profile.

Cesium-137 was detected in throughfall and stemflow samples collected from cedar and mixed broad-leaved forest stands (Fig. 2.11b, c). Cesium-137 concentrations showed significant variability among the different sampling periods; however, relatively high concentrations exceeding 10 Bq L^{-1} were observed during the early phase of the initial fallout input. The ^{137}Cs concentration in the throughfall exponentially decreased, and the rapid decrease during the early phase of the accident was followed by a slower decrease. However, a double exponential decreasing trend of ^{137}Cs concentration was masked by a significant variability among the different sampling periods such as the seasonal effects of tree physiology. The trend of decreasing ^{137}Cs concentrations in throughfall was approximated separately for the earlier (<2 years) and latter period (>2 years) of the accident. The environmental half-life for the latter period was calculated for further discussions.

The environmental half-life was calculated for the latter period of the accident (dataset of >2 years after the accident), for the throughfall was 1.3 years for the mature cedar, 1.9 years for the young cedar, and 2.1 years for the mixed broad-leaved forests. In the cedar stands, the environmental half-life for the throughfall was comparable or slightly faster than those for the old/new needles (2.0–2.2 years for the mature cedar and 2.4–3.0 years for the young cedar, $p > 0.05$). In contrast, in the mixed broad-leaved forest, the environmental half-life of the measured ^{137}Cs concentration in the throughfall (2.2 years) was markedly rapid compared to the constant concentrations found in the new leaves of Japanese konara oak (6.3 years).

Cesium-137 concentrations in the foliage/bark, litterfall, and hydrological components during each sampling year were compared to clarify the key processes involved in ^{137}Cs transfer and leaching from the tree bodies to rainwater in the studied forest site. Mean ^{137}Cs concentrations during each sampling year were calculated for litterfall, throughfall, stemflow, needles/leaves, and outer bark samples (Fig. 2.12).

Temporal changes in the ^{137}Cs concentration ratio of the litterfall to old cedar needles/new konara oak leaves are shown in Fig. 2.12a. Notably, the data for the very early phase of the accident (<200 days) are separately plotted. In the cedar stands, the ^{137}Cs concentration ratio of the litterfall to old needles tended to linearly decrease with time. However, in the mixed broad-leaved forest, the ^{137}Cs concentration ratio was higher than that of the cedar stands during the early phase of the accident (<200 days after the accident), and the concentration ratio exponentially decreased with time during the following period.

Temporal changes in the ^{137}Cs concentration ratio of throughfall to old cedar needles/new konara oak leaves are shown in Fig. 2.12b. Notably, the data for the early phase of the accident (<200 days) are separately plotted. In both the cedar and mixed broad-leaved forests, the ^{137}Cs concentration ratio exponentially decreased with time (Fig. 2.12b). The ^{137}Cs concentration ratio tended to decrease over time, suggesting that the leachability of ^{137}Cs from the tree canopy to throughfall has been decreasing during the last 6 years. Furthermore, similar temporal trends in the concentration ratio were detected for both the cedar and mixed broad-leaved forest stands. For the mixed broad-leaved forest, the ^{137}Cs leaching from the pine canopies could have affected the observed decreasing trend of ^{137}Cs concentration in throughfall. However, the canopy coverage area of the konara oak accounted for more than 75% of the total plot area. Therefore, we considered that the majority of ^{137}Cs in the throughfall originated from the canopy of the konara oak trees particularly during the leaf-on season (May to late November), but this was not the case for the litterfall.

Temporal changes in the ^{137}Cs concentration ratio of stemflow to outer bark are shown in Fig. 2.12c. In the mature cedar stand, an increasing trend in the ^{137}Cs concentration ratio from 2012 to 2014 was followed by a marked decrease during 2015 and 2016. In the young cedar stand, the ^{137}Cs concentration ratio was nearly constant from 2012 to 2014, but it tended to decrease during 2015 and 2016. However, the ^{137}Cs concentration ratio for the Japanese konara oak trees showed an increasing trend during 2013–2016 with a lower ratio during 2012. These results suggest that the formation of the ^{137}Cs concentration is likely more complex in stemflow than in throughfall. However, we only collected outer bark samples from breast height, and further investigation is required to clarify the mechanisms of ^{137}Cs entrainment from the outer bark surface to stemflow.

Cumulative ^{137}Cs deposition fluxes on the forest floor were calculated based on water inputs via throughfall and stemflow, litterfall amount, and the ^{137}Cs concentrations in those samples. Total ^{137}Cs deposition fluxes during the study period for the mature cedar, young cedar, and mixed broad-leaved stands are shown in Table 2.5, together with the contributions of throughfall, stemflow, and litterfall to the total fluxes. In the mature and young cedar stands, more than half of the total

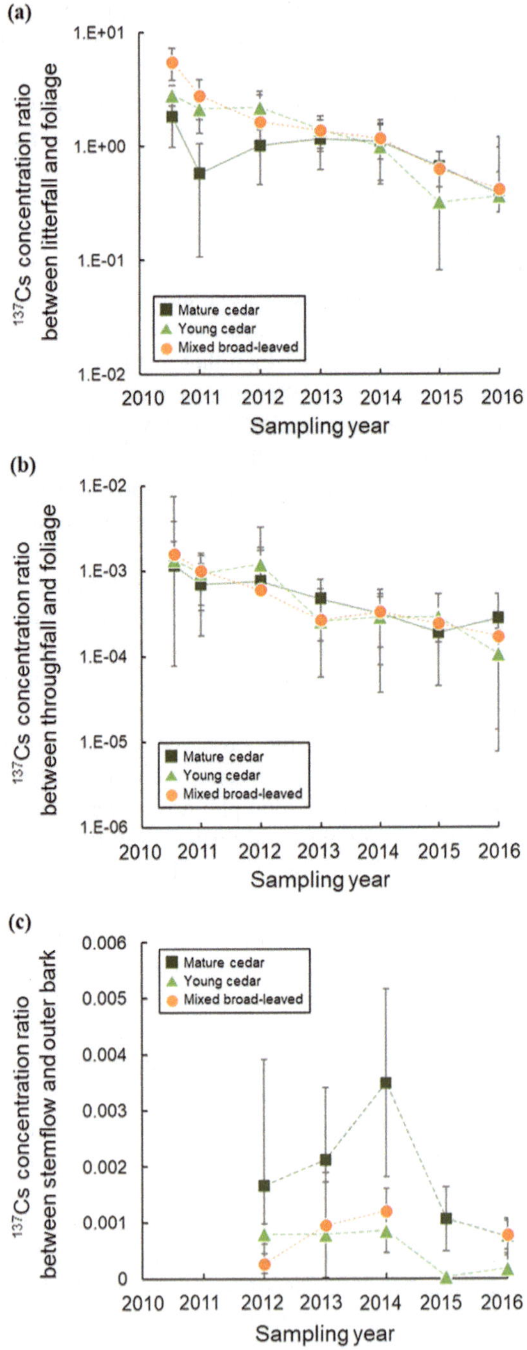

Fig. 2.12 Temporal change in the ^{137}Cs concentration ratio of litterfall/hydrological samples to vegetal sample: (**a**) litterfall and foliage, (**b**) throughfall and foliage, and (**c**) stemflow and outer bark. The error bar represents the propagation of uncertainties for the 137Cs concentrations of the paired samples

Table 2.5 Yearly ^{137}Cs depositional flux onto forest floor (loss of canopy inventory)

	Mature cedar				Young cedar				Mixed broad-leaved			
		Contribution ratio (%)				Contribution ratio (%)				Contribution ratio (%)		
	Total deposition (kBq/m^2)	TF	SF	LF	Total deposition (kBq/m^2)	TF	SF	LF	Total deposition (kBq/m^2)	TF	SF	LF
2011	70	69.5	0.4	30.1	45	56.3	2.1	41.6	24	31.4	3.7	64.9
2012	49	44.7	1.5	53.8	61	59.4	2.7	37.9	18	31.2	6.0	62.8
2013	36	17.5	0.4	82.1	61	8.1	1.0	90.9	12	14.4	4.8	80.8
2014	13	38.5	4.4	57.1	15	49.4	6.6	44.0	7	54.4	5.3	40.2
2015	15	28.5	0.4	71.1	16	45.8	0.3	53.8	3	99.4	<0.1	0.6
2016	7	47.1	0.7	52.2	6	21.8	1.2	77.0	4	48.3	3.8	47.9
2017	5	21.1	0.8	78.2	8	12.2	0.6	87.2	5	37.5	4.6	57.9
Total	196	46.4	1.0	52.6	212	39.3	2.1	58.7	72	35.0	4.5	60.5

deposition flux occurred via throughfall, and the rest was mostly via litterfall. The mixed broad-leaved stand had the greatest deposition flux, with more than half of the total deposition occurring via litterfall and most of the remaining via throughfall. The ^{137}Cs deposition flux via stemflow was less than 5% of the total in all forest stands.

The ^{137}Cs deposition fluxes via throughfall, stemflow, and litterfall showed considerable temporal variability (Fig. 2.13). In the cedar stands, throughfall was the major pathway of ^{137}Cs as it transferred from the canopy to forest floor within 200 days of the reactor accident (before mid-September 2011). We detected an increase in the ^{137}Cs deposition fluxes via throughfall around 350 days after the initial fallout. However, the contribution of litterfall to the transfer of ^{137}Cs increased over time after October 2011 (approximately 200 days after the accident). In contrast, in the mixed broad-leaved stand, litterfall made a greater contribution to ^{137}Cs transfer, and the total ^{137}Cs deposition flux via litterfall exceeded the fluxes via throughfall and stemflow.

Temporal changes in the canopy inventory of radionuclides can be expressed by the double exponential field-loss models that assume the loss of radionuclides via weathering processes, such as rain and wind (e.g., Ertel et al. 1989; Kinnersley et al. 1996; Madoz-Escande et al. 2005). We derived temporal changes in the canopy inventory for the study sites using the observed 137Cs deposition flux via throughfall, stemflow, and litterfall. The double exponential model can be expressed as follows (e.g., Kinnersley et al. 1996; Madoz-Escande et al. 2005):

$$A(t) = A_1 e^{-\lambda_1 t} + A_2 e^{-\lambda_2 t} \tag{2.5}$$

where A_1 and A_2 denote the initial interception levels of materials lost rapidly and slowly, respectively, and λ_f and λ_s represent the rate constants for the rapid and slow loss rate (year^{-1}) of the canopy inventory, respectively.

The canopy ^{137}Cs inventory for the three forest stands is plotted against time (Fig. 2.14, Table 2.6). The figure shows the loss of canopy-intercepted ^{137}Cs as a result of throughfall, stemflow, and litterfall plotted against time after the reactor accident. Additionally, data were collected from previous studies (Kato et al. 2012; Teramage et al. 2014) to complement lack of observation data during the very early period of the accident. The loss of ^{137}Cs from various forest canopies, as derived from the ^{137}Cs transfer measured in the forest stands, was determined using a double exponential field-loss model (Eq. 2.5).

The double exponential model fitted the observed data with high levels of significance for all of the forest stands (Fig. 2.14). The canopy ^{137}Cs inventory has been directly measured by cutting nine standing cedar trees and collecting organic and mineral soil layer from the forest floor in November 2013 (Coppin et al. 2016). The resulting ^{137}Cs inventory in the aboveground tree biomass for the mature and young cedar was 14.0% and 15.8% of total atmospheric deposition input of ^{137}Cs following the Fukushima accident. These measured ^{137}Cs canopy inventories were

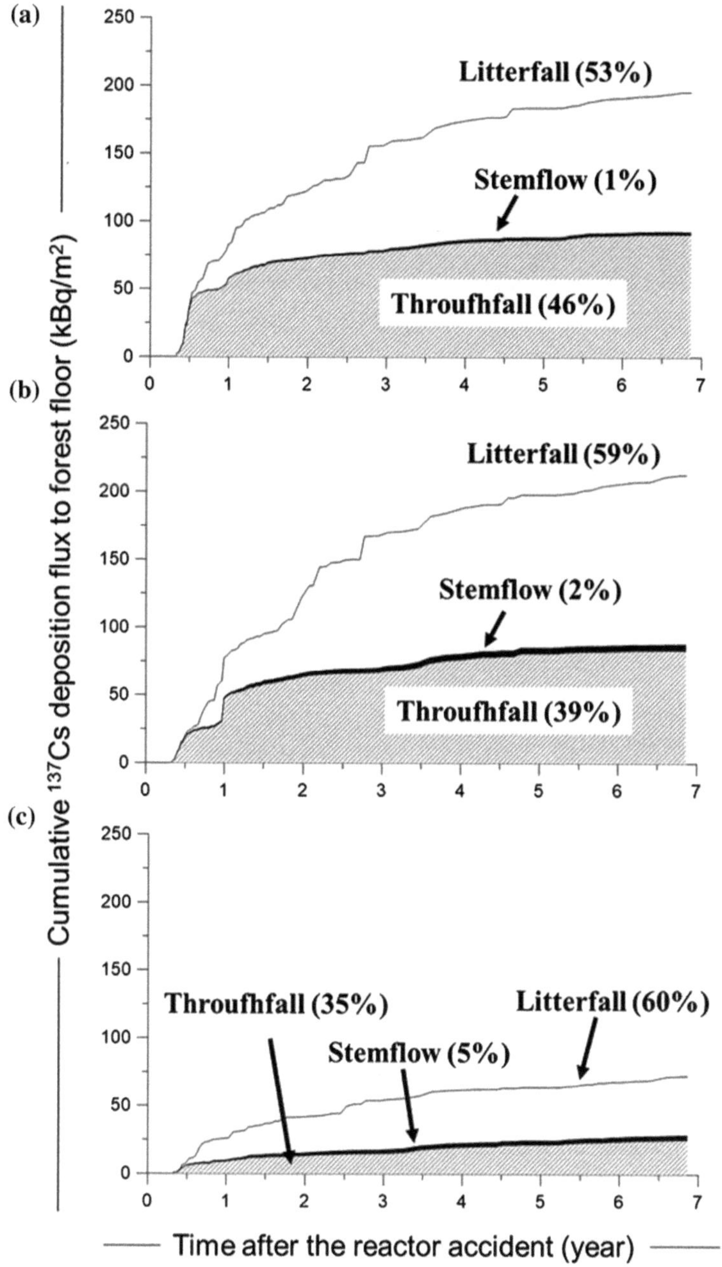

Fig. 2.13 Cumulative ^{137}Cs deposition onto forest floor via various pathways during the 6-year observation period. (**a**) Mature ceder, (**b**) young ceder, (**c**) mixed broad-leaved

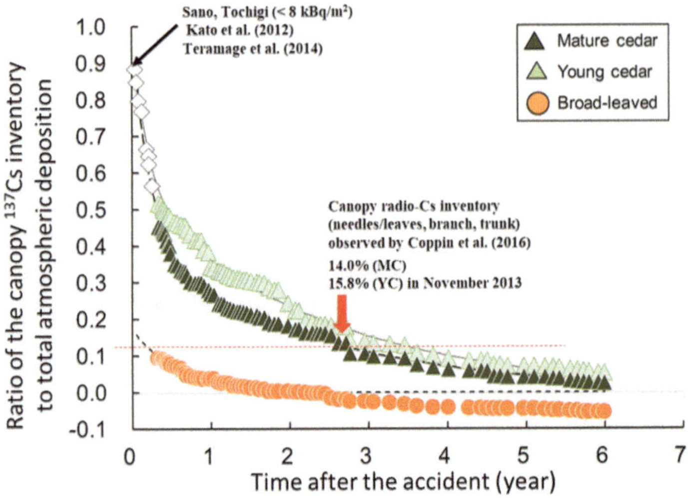

Fig. 2.14 Temporal change in the ^{137}Cs concentration ratio of litterfall/hydrological samples to vegetal sample: (**a**) litterfall and foliage, (**b**) throughfall and foliage, and (**c**) stemflow and outer bark. The error bar represents the propagation of uncertainties for the 137Cs concentrations of the paired samples

Table 2.6 Parameters for double exponential field-loss model

	MC	YC	BL
A_1	0.60	0.45	0.12
A_2	0.40	0.55	0.05
λ_f (year^{-1})	4.44	5.11	1.69
λ_s (year^{-1})	0.43	0.42	1.69
$T_{f1/2}$ (year)	0.16	0.14	0.41
$T_{s1/2}$ (year)	1.61	1.66	0.41

highly comparable with those estimated by the double exponential field-loss model in this study.

The initial canopy interception of deposited radionuclides in the forest stands studied here was estimated to be 100% for both the mature and young cedar stands and 17% for the mixed broad-leaved stand. An initial canopy interception rate of >90% was reported for radiocesium in a 41-year-old Japanese cedar stand (*Cryptomeria japonica*) with a stand density of 1300 ha^{-1} and a 40-year-old cypress stand (*Chamaecyparis obtusa*) with a stand density of 2500 ha^{-1} (Kato et al. 2012). Ronneau et al. (1987) studied the deposition of Chernobyl-derived ^{137}Cs onto a pine forest in the east of Belgium. They reported a very effective retention of the deposited ^{137}Cs by plant foliage, and the initial ^{137}Cs interception by pine canopies was estimated to be 79% of the atmospheric input. An 85-year-old Norway spruce stand in Munich, Germany, with a stand density of 622 ha^{-1} had an initial ^{137}Cs interception rate of 70% for an atmospheric deposition of 20 kBq m-2 during April

27–30, 1986 (Bunzl et al. 1989). Melin et al. (1994) studied the retention and distribution of radionuclides in Swedish coniferous (an old pine forest with a stem density of 1622 ha^{-1}) and deciduous forests (an approximate stem density of 600 ha^{-1}) immediately after the Chernobyl accident. Their rough estimate indicated that a spruce stand will intercept about 90% of the deposited radionuclides, while unfoliated deciduous stands of beech, birch, and alder will intercept less than 35% of the dry deposited radionuclides. The estimated initial canopy interception for the forests studied here was comparable with values reported in the contaminated forests affected by the Chernobyl and Fukushima accidents (Tables 2.2 and 2.3).

2.5 Summary

A wide range of forest environments have been contaminated by the atmospherically deposited radiocesium following the Fukushima Daiichi Nuclear Power Plant accident. The forest area accumulated in total is 1.8 PBq of ^{137}Cs, which is corresponding to 72% of total ^{137}Cs activities deposited on the land area of Japan. The contaminated forest except for the vicinity of residential area has not been subjected to decontamination so far; it still holds most of the ^{137}Cs deposited during the initial phase of the accident. However, radiocesium is highly mobile within forest ecosystems; therefore temporal evolution of its distributions and transfer among different compartments should be observed and modeled in order to predict the future fate of the accidentally deposited radiocesium in temperate forest environment.

Canopy interception of atmospherically deposited radiocesium varied significantly among different tree species. The evergreen conifers tend to show high canopy interception rate greater than 70% of atmospheric input in most cases. The high interception fraction of the deposited radiocesium by coniferous canopies indicated that the canopy will act as a secondary source of radioactive contamination of the forest floor. On the other hand, deciduous broad-leaved showed relatively lower interception rate less than 40%. These observations were likely common through the forests affected by the Chernobyl and Fukushima accidents although there were significant differences in physicochemical form of initial fallout and conditions of climate between those accidents. Nevertheless, the observed interception rate for deciduous forest species represented initial canopy interception during leaf-off season of the year; canopy interception by deciduous broad-leaved forest has not been sufficiently clarified because there have been limited data available for the canopy interception for deciduous broad-leaved species during growing season. Furthermore, initial canopy interception rate varied significantly even for the same tree species; further investigation is necessary to clarify influences of stand properties and deposition type (e.g., wet or dry dominated deposition) on the efficiency of initial canopy interception.

Canopy interception of atmospherically deposited radiocesium is followed by hydrological and biological self-decontamination processes. Temporal trend of canopy radiocesium removal is important for determining temporal evolution of contamination of forest soils and subsequent cycling of radiocesium within forest ecosystems. The monitoring of radiocesium concentrations in hydrological and biological components was effective to determine the transport of canopy radiocesium inventory to forest floor. Furthermore, a double exponential field-loss model can successfully simulate the observed loss of canopy radiocesium from Japanese cedar and konara oak forest mixed with red pine during the early phase of the accident. However, further investigation is necessary to clarify the mechanisms of radiocesium leaching from the forest canopies and the effects of root uptake prevailing long-term radiocesium migration from soil profile to tree under quasi-equilibrium condition.

References

Amano H, Akiyama M, Chunlei B, Kawamura T, Kishimoto T, Kuroda T, Muroi T, Odaira T, Ohta T, Takeda K, Watanabe Y, Morimoto T (2012) Radiation measurements in the Chiba Metropolitan Area and radiological aspects of fallout from the Fukushima Dai-ichi Nuclear Power Plants accident. J Environ Radioact 111:42–52

Bunzl K, Schimmack W, Kreutzer K, Schierl R (1989) Interception and retention of Chernobyl-derived ^{134}Cs, ^{137}Cs and ^{106}Ru in a spruce stand. Sci Total Environ 78:77–87. https://doi.org/10.1016/0048-9697(89)90023-5

Butler C (2011) Radioactivity spreads in Japan. Nature 471:555–556

Chino M, Nakayama H, Nagai H, Terada H, Katata G, Yamazawa H (2011) Preliminary estimation of release amounts of ^{131}I and ^{137}Cs accidentally discharged from the Fukushima Daiichi nuclear power plant into the atmosphere. J Nucl Sci Technol 48(7):1129–1134

Coppin F, Hurtevent P, Loffredo N, Simonucci C, Julien A, Gonze M-A, Nanba K, Onda Y, Thiry Y (2016) Radiocaesium partitioning in Japanese cedar forests following the "early" phase of Fukushima fallout redistribution. Sci Rep 6(1)

Ertel J, Voigt G, Paretzke HG (1989) Weathering of 134/137Cs following leaf contamination of grass cultures in an outdoor experiment. Radiat Environ Biophys 28(4):319–326

Fukushima Prefecture (2016) Current situation and future prediction of radionuclides in forest environment. http://www.pref.fukushima.lg.jp/uploaded/attachment/221567.pdf. Last accessed on 1 July 2017

Hashimoto S, Ugawa S, Nanko K, Shichi K (2012) The total amounts of radioactivity contaminated materials in forests in Fukushima, Japan. Sci Rep 2:416. https://doi.org/10.1038/srep00416

Hashimoto S, Matsuura T, Nanko K, Linkov I, Shaw G, Kaneko S (2013) Predicted spatio-temporal dynamics of radiocesium deposited onto forests following the Fukushima nuclear accident. Sci Rep 3:2564. https://doi.org/10.1038/srep02564

Hirose K (2012) Fukushima Dai-ichi nuclear power plant accident: summary of regional radioactive deposition monitoring results. J Environ Radioact 111:13–17

Hisadome K, Onda Y, Kawamori A, Kato H (2013) Migration of radiocaesium with litterfall in hardwood-Japanese red pine mixed forest and sugi plantation. J Jpn For Soc 95:267–274

IAEA (2006) Environmental consequences of the Chernobyl accident and their remediation: twenty years of experience. Report of the Chernobyl forum expert group 'Environment', Radiological Assessment Reports Series, International Atomic Energy Agency: Vienna, Austria. p 166

Imamura N, Komatsu M, Ohashi S, Hashimoto S, Kajimoto T, Kaneko S, Takano T (2017) Temporal changes in the radiocesium distribution in forests over the five years after the

Fukushima Daiichi Nuclear Power Plant accident. Sci Rep 7:8179. https://doi.org/10.1038/s41598-017-08261-x

Itoh Y, Imaya A, Kobayashi M (2015) Initial radiocesium deposition on forest ecosystems surrounding the Tokyo metropolitan area due to the Fukushima Daiichi Nuclear Power Plant accident. Hydrol Res Lett 9(1):1–7. https://doi.org/10.3178/hrl.9.1

Katata G, Chino M, Kobayashi T, Terada H, Ota M, Nagai H, Kajino M, Draxler R, Hort MC, Malo A, Torii T, Sanada Y (2015) Detailed source term estimation of the atmospheric release for the Fukushima Daiichi Nuclear Power Station accident by coupling simulations of an atmospheric dispersion model with an improved deposition scheme and oceanic dispersion model. Atmos Chem Phys 15:1029–1070. https://doi.org/10.5194/acp-15-1029-2015

Kato H, Onda Y (2018) Determining the initial Fukushima reactor accident-derived cesium-137 fallout in forested areas of municipalities in Fukushima Prefecture. J For Res 23:73–84. https://doi.org/10.1080/13416979.2018.1448566

Kato H, Onda Y, Gomi T (2012) Interception of the Fukushima reactor accident-derived 137Cs, ^{134}Cs and ^{131}I by coniferous forest canopies. Geophys Res Lett 39:L20403. https://doi.org/10.1029/2012GL052928

Kato H, Onda Y, Loffredo N, Hisadome K, Kawamori A (2017) Temporal changes in radiocesium deposition in various forest stands following the Fukushima Dai-ichi Nuclear Power Plant accident. J Environ Radioact 116(3):449–457. https://doi.org/10.1016/j.jenvrad.2015.04.016

Kato H, Onda Y, Gao X, Sanada Y, Saito K (2019) Reconstruction of a Fukushima accident-derived radiocesium fallout map for environmental transfer studies. J Environ Radioact

Kinnersley RP, Shaw G, Bell JNB, Minski J, Goddard AJH (1996) Loss of particulate contaminants from plant canopies under wet and dry conditions. Environ Pollut 91(2):227–235. https://doi.org/10.1016/0269-7491(95)00047-X

Kitamura A, Yamaguchi M, Kurikami H, Yui M, Onishi Y (2014) Predicting sediment and cesium-137 discharge from catchments in eastern Fukushima. Anthropocene 5:22–31

Madoz-Escande C, Garcia-Sanchez L, Bonhomme T, Morello M (2005) Influence of rainfall characteristics on elimination of aerosols of cesium, strontium, barium and tellurium deposited on grassland. J Environ Radioact 84:1–20

Melin J, Wallberg L, Suomela J (1994) Distribution and retention of cesium and strontium in Swedish boreal forest ecosystems. Sci Total Environ 157:93–105

MEXT (2011a) Preparation of distribution map of radiation doses, etc. (map of radioactive cesium concentration in soil) by MEXT. radioactivity.nsr.go.jp/en/contents/5000/4165/24/1750_083014.pdf. Last accessed on 1 Jul 2017

MEXT (2011b) Results of the third airborne monitoring survey by MEXT. radioactivity.nsr.go.jp/en/contents/5000/4182/24/1304797_0708e.pdf. Last accessed on 1 July 2017

Ministry of Agriculture, Forestry and Fisheries of Japan (MAFF) (2011) Monitoring results of ambient dose rate in forest of Fukushima prefecture. In Japanese. http://www.rinya.maff.go.jp/j/press/hozen/111227_3.html (Last accessed on July 1, 2017)

Niinemets Y, Tamm U (2005) Species differences in timing of leaf fall and foliage chemistry modify nutrient resorption efficiency in deciduous temperate forest stands. Tree Physiol 25:1001–1014

Nuclear Regulation Authority of Japan (NRA) (2017) Monitoring information of environmental radioactivity level. Monitoring survey on radionuclide distribution in environment. Available at http://radioactivity.nsr.go.jp/ja/list/338/list-1.html. Last accessed 1 July 2017. In Japanese

Okada N, Nakai W, Ohashi S, Tanaka A (2015) Radiocesium migration from the canopy to the forest floor in pine and deciduous forests. J Jpn For Soc 97:57–62

Onda Y, Kato H, Hoshi M, Takahashi Y, Nguyen M-L (2015) Soil sampling and analytical strategies for mapping fallout in nuclear emergencies based on the Fukushima Dai-ichi nuclear power plant accident. J Environ Radioact 139:300–307

Rauret G, Llauradó M, Tent J, Rigol A, Alegre LH, Utrillas MJ (1994) Deposition on holm oak leaf surfaces of accidentally released radionuclides. Sci Total Environ 157:7–16

Ronneau C, Cara J, Apers D (1987) The deposition of radionuclides from Chernobyl to a forest in Belgium. Atmos Environ 21(6):1467–1468

Schimmack W, Bunzl K, Kreutzer K, Rondenkirchen E, Schierl R (1991) Einfluss von fichte (*Picea abies L. Karst*) und buche (*Fagus sylvatica* L.) auf die Wanderung von radiocasium im Boden. Fortwiss Forsch 39:242e251

Sombre L, Vanhouche M, Thiry Y, Ronneau C, Lambotte JM, Myttenaere C (1990) Transfer of radiocesium in forest ecosystems resulting from a nuclear accident. In: Desmet G et al (eds) *Transfer of radionuclides in natural and semi-natural environments*. Elsevier Applied Science, p 74e83

Teramage MT, Onda Y, Kato H, Gomi T (2014) The role of litterfall in transferring Fukushima-derived radiocesium to a coniferous forest floor. Sci Total Environ 490:435–439

Thiry Y (1997) Etude du cycle du radiocesium en ecosysteme forestier: Distribution et facteurs de mobilité. Thesis, Université Catholique de Louvain, Louvain-la-Neuve, Belgium

Thiry Y, Garcia-Sanchez L, Hurtevent P (2016) Experimental quantification of radiocesium recycling in a coniferous tree after aerial contamination: field loss dynamics, translocation and final partitioning. J Environ Radioact 161:42–50

Tikhomirov FA, Shcheglov AI (1991) The radiological consequences of the Kyshtym and Chernobyl radiation accidents for forest ecosystems. In: Proceedings of the Seminar on Comparative Assessment of the Environmental Impact of Radionuclides Released during Three Major Nuclear Accidents, Kyshtym, Windscale, Chernobyl. CEC, Luxembourg, 1-5 October 1990, H, EUR 13574, pp 867–888

Part II
Mechanisms of Radiocesium Translocation in Plants

Chapter 3
Uptake of Radiocesium by Plants

Yuki Sugiura and Chisato Takenaka

Abstract Radiocesium transfer to plants immediately after the Fukushima Daiichi Nuclear Power Plant (FDNPP) accident was investigated by collecting soil and newly emerged leaf samples between May 2011 and November 2012 from 20 sites in the Fukushima Prefecture. Autoradiograph images indicated that woody plants tend to accumulate radiocesium in their newly emerged leaves than herbaceous plants due to translocation-deposited radiocesium on old stems and/or leaves. Radiocesium uptake by plants was evaluated by ratios of the radiocesium concentration in plant to that in soil (concentration ratio, CR). Although the CR values decreased in 2012, CR values of woody plants remain high compared to that of herbaceous plants. Exchangeable ^{137}Cs rates in soil (extraction rate) were measured at five sites, and these rates decreased at four sites in 2012 and depended on environmental conditions and soil type. Both CR values and extraction rates decreased in 2012; however, CR values reflected the changes in not only extraction rates but also characteristics of each species. Amaranthaceae, Chenopodiaceae, and Polygonaceae, which had been identified as Cs accumulators, presented no clear ^{137}Cs accumulation ability. In 2012, the perennial plant *Houttuynia cordata* and deciduous trees *Chengiopanax sciadophylloides* and *Acer crataegifolium* displayed high CR values, indicating that these species are ^{137}Cs accumulators.

Keywords Radiocesium · Fukushima Daiichi Nuclear Power Plant accident · Radiocesium translocation · Concentration ratio · Cs accumulator

---

Y. Sugiura (✉)
Graduate School of Bioagricultural Sciences, Nagoya University, Nagoya, Aichi, Japan

Nuclear Fuel Cycle Engineering Laboratories, Japan Atomic Energy Agency, Tokai, Japan
e-mail: sugiura.yuki@jaea.go.jp

C. Takenaka
Graduate School of Bioagricultural Sciences, Nagoya University, Nagoya, Aichi, Japan

C. Takenaka et al. (eds.), *Radiocesium Dynamics in a Japanese Forest Ecosystem*,
https://doi.org/10.1007/978-981-13-8606-0_3

3.1 Introduction

The Great East Japan Earthquake and the succeeding tsunami caused serious damages to the Fukushima Daiichi Nuclear Power Plant (FDNPP) on March 11, 2011. Numerous radionuclides were released to the environment through venting and explosion. Main radionuclides released from FDNPP were radiocesium (^{137}Cs, ^{134}Cs) and radioiodine (^{131}I), and the total amounts of ^{137}Cs and ^{131}I were estimated to be approximately 1.3×10^{16} Bq and 2.0×10^{17} Bq, respectively (Chino et al. 2011; Kobayashi et al. 2013; Hirose 2016). Half-lives of ^{134}Cs, ^{137}Cs, and ^{131}I are 2.1 years, 30 years, and 8.1 days, respectively. Radioiodine has not been detected in the environment after several years passed from the accident. On the other hand, since ^{137}Cs has relatively long half-life, its impacts on humans and ecosystems are concerns. Radiocesium was dispersed and deposited over a wide area of Fukushima Prefecture, especially at northwest of FDNPP (Chino et al. 2011; NRA 2014). Approximately 70% area of Fukushima Prefecture is covered with forests (Forest Agency 2014), and a large area of forest has been contaminated with radiocesium (Kitamura et al. 2014). Some part of radiocesium deposited onto forest would be translocated by rainfall and wind and finally reach ocean area and living sphere (Iijima 2015; JAEA 2015). Thus, understanding the dynamics of radiocesium in terrestrial, especially forest, ecosystem and decontamination are crucial issues.

Radiocesium migration from soil to plants has been investigated after the Chernobyl Nuclear Power Plant (ChNPP) accident mainly in Europe. According to the researches, radiocesium accumulation abilities vary by plant species, and some species of Amaranthaceae (Broadley et al. 1999; Dushenkov et al. 1999; Fuhrmann et al. 2003; Lasat et al. 1998), Chenopodiaceae, and Polygonaceae (Broadley and Willey 1997) are reported as potential ^{137}Cs accumulators. In addition, soil characteristics also important factor for ^{137}Cs uptake (Ehlken and Kirchner 2002; Tsukada and Hasegawa 2002).

Cesium uptake by plants had also been investigated in Japan before the FDNPP accident, particularly for the stable isotope ^{133}Cs and low-concentration ^{137}Cs derived from global fallout and the ChNPP accident (Kamei-Ishikawa et al. 2008; Tsukada and Nakamura 1999). However, the behavior of ^{137}Cs just after a substantial deposition on the terrestrial ecosystem is unclear. Because the Japanese islands are located in the most active part of the Pacific Ring of Fire, andosol, a soil derived from volcanic ash, is widely distributed in Japan. This rare soil exhibits unique properties, such as low bulk density, high water-holding capacity, and high phosphate-fixing capacity (Takeda et al. 2013). Also, land use, climate, and rainfall patterns differ between Japan and Europe. Therefore, the knowledge obtained from investigations on the ChNPP accident needs to be carefully applied to the FDNPP accident (Yamaguchi et al. 2012). A continuous observation of the dynamics of ^{137}Cs immediately after the FDNPP accident is crucial.

The ^{137}Cs contamination of agricultural products has been studied and monitored directly after the FDNPP accident, but wild plants have attracted less attention (Mimura et al. 2014). Nonetheless, primary information collected about ^{137}Cs transfer for many plants may help with the decontamination of agricultural plants and handling of contaminated weeds (Yamashita et al. 2014). Moreover, identified ^{137}Cs accumulators would be applied to phytoremediation.

In the present study, we collected plant and soil samples in Fukushima Prefecture from May 2011 to November 2012 to obtain primary information on the characteristics of ^{137}Cs accumulation immediately after the accident.

3.2 Cesium-137 Concentrations in Soil and Extraction Rates

Sampling sites and their land use are described in Fig. 3.1 and Table 3.1, respectively. Twenty sites in seven municipalities of Fukushima Prefecture were surveyed at least once per year to evaluate yearly changes. Samples were also collected irregularly from five sites located in three municipalities.

Cesium-137 concentrations in soil for each sampling site are tabulated in Table 3.2. The values varied widely across sites. The extraction rates of the five sites are shown in Fig. 3.2. In the present study, extraction rates were calculated as follows:

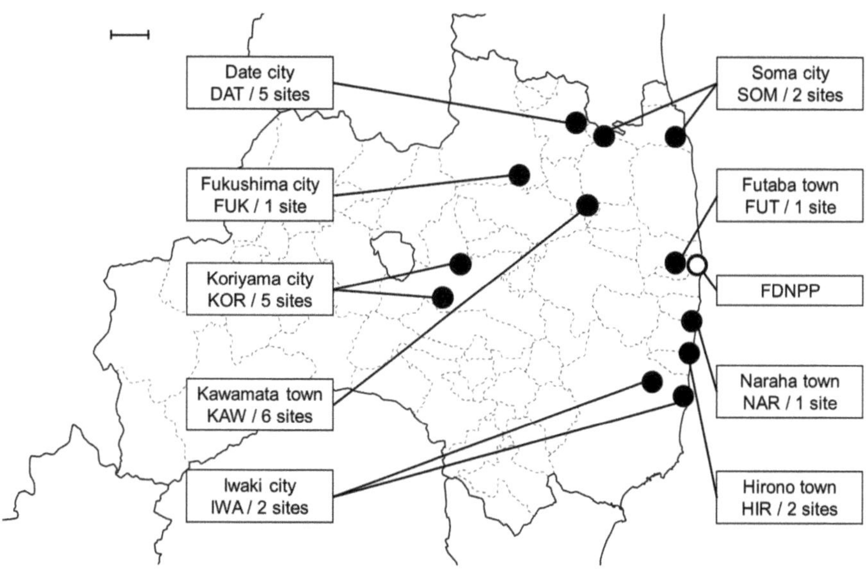

Fig. 3.1 Map of the sampling sites and FDNPP. The bar indicates 10 km

Table 3.1 Land use, soil taxonomy, and soil texture of each sampling site

Municipality	Sampling site	Land use (vegetation)	Soil type	Soil texture[a]	Yearly changes[b]	Exchangeable ^{137}Cs[b]
Date city	DAT1	Broadleaved forest, grassland	Brown forest soil	L, SiL	○	
	DAT2	Broadleaved forest, grassland	Brown forest soil	L, SiL	○	○
	DAT3	Broadleaved forest	Brown forest soil	L, SiL	○	
	DAT4	Paddy field	Andosol	L, SiL	○	
	DAT5	Broadleaved forest, grassland	Andosol	L, SiL	○	
Fukushima city	FUK	Bamboo forest	Brown forest soil	CL, SiCL, SCL		
Futaba town	FUT	Grassland, field	Andosol	L, SiL	○	
Hirono town	HIR1	Grassland	Brown lowland soil	SL	○	
	HIR2	Grassland	Andosol	SL		
Iwaki city	IWA1	Grassland	Brown lowland soil	L, SiL	○	
	IWA2	Broadleaved forest	Red-yellow soil	L, SiL	○	
Kawamata town	KAW1	Broadleaved forest	Brown forest soil	L, SiL	○	
	KAW2	Field	Andosol	CL, SiCL, SCL	○	
	KAW3	Meadow	Brown lowland soil	L, SiL	○	
	KAW4	Broadleaved forest, grassland	Brown forest soil	L, SiL		
	KAW5	Broadleaved forest, grassland	Brown forest soil	L, SiL		
	KAW6	Broadleaved forest	Brown forest soil	L, SiL		
Koriyama city	KOR1	Field	Andosol	CL, SiCL, SCL	○	○

(continued)

Table 3.1 (continued)

Municipality	Sampling site	Land use (vegetation)	Soil type	Soil texture[a]	Yearly changes[b]	Exchangeable ^{137}Cs[b]
	KOR2	Grassland	Andosol	L, SiL	○	○
	KOR3	Broadleaved forest, grassland	Brown forest soil	L, SiL	○	○
	KOR4	Broadleaved forest	Brown forest soil	L, SiL	○	
	KOR5	Broadleaved forest	Brown forest soil	L, SiL	○	
Naraha town	NAR	Mixed forests of conifers and broadleaved trees	Red-yellow soil	L, SiL	○	○
Soma city	SOM1	Grassland	Andosol	SL	○	
	SOM2	Grassland	Brown lowland soil	S, LS	○	

[a]*S* sand, *SL* sandy loam, *L* loam, *SiL* silt loam, *SCL* sandy clay loam, *CL* clay loam, *SiCL* silty clay loam
[b]○ measured

$$\text{Extraction rate }(\%) = \frac{\text{Amount of extracted }^{137}\text{Cs (Bq)}}{\text{Amount of }^{137}\text{Cs in soil (Bq)}} \times 100$$

The significance of differences in the extraction rate between 2011 and 2012 was assessed by the Steel-Dwass test ($p < 0.05$). The yearly change in extraction rate for this period at the same site was expressed as the ratio between the extraction rates obtained in 2012 (%) and 2011 (%). The extraction rates decreased in 2012 except for site NAR1, which displayed the highest rate (25 ± 7.1%) in 2011 and maintained a high value in 2012 (31 ± 1.1%). Site KOR1 exhibited minimum extraction rates that amounted to 2.3 ± 0.1% and 0.7 ± 0.1% in 2011 and 2012, respectively. Significant ($p < 0.05$) yearly changes were detected at sites KOR1 and DAT2.

It is inferred that the decrease of extraction rate is attributed to ^{137}Cs sorption and fixation by clay minerals, which limits mobility over time after deposition onto soil (Takeda et al. 2013). The extraction rates differed among sampling sites. Soil organic matter and clay minerals act as ligands for Cs. In particular, 2:1 type layer silicates such as mica, illite, and vermiculite retain Cs more strongly than other ligands (Yamaguchi et al. 2012). Therefore, ^{137}Cs sorbed by these clay minerals would be difficult to extract using 1 M ammonium acetate solution. Sites KOR1 and KOR2 were used as field and grassland, respectively, and were consist of largely

Table 3.2 Radiocesium concentrations in soil for each sampling site

Sampling site	Replicate	Mean	±	S.E.	Maximum	Minimum	Median
DAT1	14	5000	±	1320	22,900	522	2940
DAT2	15	4370	±	730	10,300	516	3220
DAT3	14	5520	±	872	10,100	688	5930
DAT4	6	3940	±	463	5260	2070	4290
DAT5	13	5110	±	1340	62,600	586	4750
FUK	3	1720	±	625	2740	586	1820
FUT	6	77,100	±	16,500	131,000	26,700	90,400
HIR1	9	80	±	6	1830	N.D.	95
HIR2	1	8940	±	–	–	–	–
IWA1	4	400	±	175	870	37	347
IWA2	14	2140	±	351	4810	670	2010
KAW1	8	2140	±	351	4810	670	2010
KAW2	9	7160	±	986	12,200	2330	7700
KAW3	6	16,000	±	5070	35,400	3780	13,700
KAW4	1	3240	±	–	–	–	–
KAW5	1	2700	±	–	–	–	–
KAW6	1	4530	±	–	–	–	–
KOR1	16	6370	±	848	11,900	930	6560
KOR2	17	5880	±	1680	30,500	953	4250
KOR3	20	4060	±	543	15,000	734	3810
KOR4	13	1220	±	183	2540	462	987
KOR5	15	1870	±	361	4550	226	1470
NAR	17	2990	±	535	8300	597	2390
SOM1	15	798	±	24	14,600	342	1130
SOM2	5	118	±	26	430	45	149

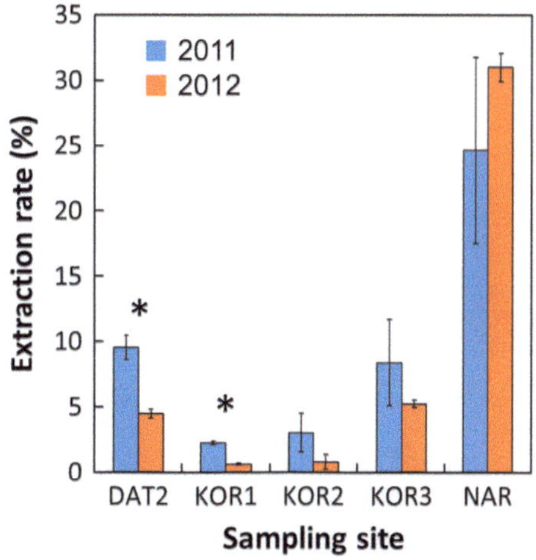

Fig. 3.2 Mean extraction rates at five sites; $n = 3$, error bar = S.E., *: $p < 0.05$

clayey soils. In contrast, the other sites were forests and the soils contain rich organic matter. These soil components affected the difference of exchangeable ^{137}Cs rates among the sites. Soil collected from site NAR1 was red-yellow soil, and its components may explain the highest extraction ^{137}Cs rate and the dissimilar yearly trend. Further analyses, such as organic matter content and mineralogical composition, will be valid for a precise evaluation.

3.3 Differences in ^{137}Cs Accumulation According to the Plant Life Form

3.3.1 Autoradiograph Analysis

Figure 3.3-I shows an autoradiograph image after 48-h exposure of foliar samples collected at site DAT5 in May 2011. Images were obtained from the leaves of woody plants only. Some hotspots, indicative of highly intense radionuclide enrichment, were found in/on the leaves. Figure 3.3-II shows an autoradiograph image after 90-h exposure of the evergreen tree *Quercus myrsinifolia* collected at site FUK1 on March 2012. Branches and leaves present during the accident displayed sharp images with hotspots. In contrast, newly emerged branches and leaves exhibited unclear images.

The major radionuclides in plant samples, which provided autoradiograph images, are radiocesium ($^{134,\ 137}$Cs) and radiopotassium (^{40}K). Since it took over 1 week to obtain autoradiograph image of non-contaminated leaf samples (data not shown), images obtained in this study indicates the presence of radiocesium in plant tissues. For samples taken in May 2011 (Fig. 3.3-I), radiocesium was observed in/on the leaves of woody plants but not in the herbaceous plants after 48-h exposure, although all leaves grew after the accident. This result is derived from the presence or

Fig. 3.3 Spatial distribution of radionuclides in/on leaves collected at site DAT5 on May 2011 (exposed for 48 h) (I). The bar indicates 5 cm. Capital and lowercase letters correspond to woody (deciduous) and herbaceous plants (perennial), respectively. (**A**) *Castanea crenata*, (**B**) *Lespedeza bicolor*, (**C**) *Rubus palmatus* var. *coptophyllus*, and (**D**) *Cerasus jamasakura*; (**a**) *Reynoutria japonica*, (**b**) *Pteridium aquilinum*, (**c**) *Pueraria lobata*, (**d**) *Artemisia indica* var. *maximowiczii*, and (**e**) *Equisetum arvense*. Spatial distribution of radionuclides in/on the branch of the evergreen tree *Quercus myrsinifolia* collected at site FUK1 on March 2012 (exposed for 90 h) (II). The bar indicates 5 cm. Tissues that emerged after the FDNPP accident are delimited by the dashed line

absence of aboveground tissues to receive radiocesium fallout just after the accident. Woody plants absorb radiocesium through not only roots but also plant surfaces, such as stems (Tagami et al. 2012). Autoradiograph images of evergreen tree branches collected in March 2012 showed many dots on old shoots existing at the time of the accident (Fig. 3.3-II). Images of new leaves that developed at the tip of old leaves and stem after the accident indicate that these leaves contained radiocesium and have been affected by surface absorption and translocation from old leaves and stem. Similar results have been observed for *Torreya nucifera* (Sakamoto et al. 2012) and *Cryptomeria japonica* (Kanasashi et al. 2015) following the FDNPP accident. Some hotspots were observed on old and new leaves of deciduous (Fig. 3.3-I) and evergreen trees (Fig. 3.3-II), implying that a secondary dispersion of radiocesium occurred after the development of new leaves.

3.3.2 Cesium-137 Concentration Ratio Between Plant and Soil

In the present study, the "concentration ratio (CR)" was used instead of a "transfer factor (TF)" to evaluate the ^{137}Cs concentration in plants relative to that in soils because several plants may have been affected by direct deposition immediately after the accident and absorbed ^{137}Cs through the plant surface. CR was defined the same as TF:

$$\mathrm{CR} = \frac{^{137}\mathrm{Cs\ concentration\ in\ plant\ [Bq/kg\ dry\ weight\ (DW)]}}{^{137}\mathrm{Cs\ concentration\ in\ soil\ (Bq/kgDW)}}$$

The activities in plants were corrected for decay from the sampling date. Mean ^{137}Cs concentrations in soil were adjusted according to plant collection dates to calculate CR values.

Figure 3.4 shows histograms of log CR values of all plant samples collected in 2011 and 2012. In 2011, a peak of log CR value of herbaceous plants was -1.0–-0.5. Log CR value of 28 individuals are below -2 and the proportion was 21%, whereas woody plants have the peak at $-0.5 - 0$ and the proportion of individuals with log CR values below -2 was 2.6%. In 2012, the peak of log CR value of herbaceous plants shifted to -2.0–-1.5, and individuals with log CR values below -2 increased to 22%. Log CR values of woody plants also decreased; however, the change was smaller than that in herbaceous plants.

Individuals presenting the 20 highest CR values among the herbaceous or woody plants in 2011 and 2012 are listed in Table 3.3. In 2011, the breakdown of life form is 4 annual and 16 perennial herbaceous plants and 4 deciduous and 16 evergreen trees, respectively. Herbaceous and woody plants exhibiting the highest CR corresponded to *Chenopodium ficifolium* (CR = 54) and *Phyllostachys heterocycla* (CR = 30), respectively. *Houttuynia cordata* collected at multiple sites showed high CR (CR = 10, KOR4; CR = 3.3, 2.9, 2.5, KOR3). In 2012, annual and perennial herbs present in the top 20 individuals with high CR values were 5 and 15, respectively. Deciduous and evergreen trees were 16 and 4, respectively. Herbaceous and woody plants displaying the highest CR value were *Oenothera biennis* (CR = 5.0) and *Chengiopanax sciadophylloides* (*syn. Acanthopanax sciadophylloides*; *Eleutherococcus sciadophylloides*) (CR = 11), respectively. *Chengiopanax sciadophylloides* (CR = 11, KAW1; CR = 3.3, 2.3, IWA2) and *Acer crataegifolium* (CR = 11, 3.9, KAW1; CR = 1.0, IWA2) collected at multiple sites showed high CR values. Five *H. cordata* individuals, which exhibited relatively high CR in 2011, maintained high CR in 2012 (CR = 1.3, KOR4; CR = 1.2; HIR1). Several evergreen trees, such as Poaceae, Theaceae, and Ericaceae family species, presented high CR values in 2011. In contrast, this number decreased to only four evergreen trees in 2012.

Woody plants presented higher CR values than their herbaceous counterparts (Fig. 3.4 and Table 3.3). Moreover, evergreen tree leaves displayed higher CR values than their deciduous equivalents in 2011. This is consistent with findings

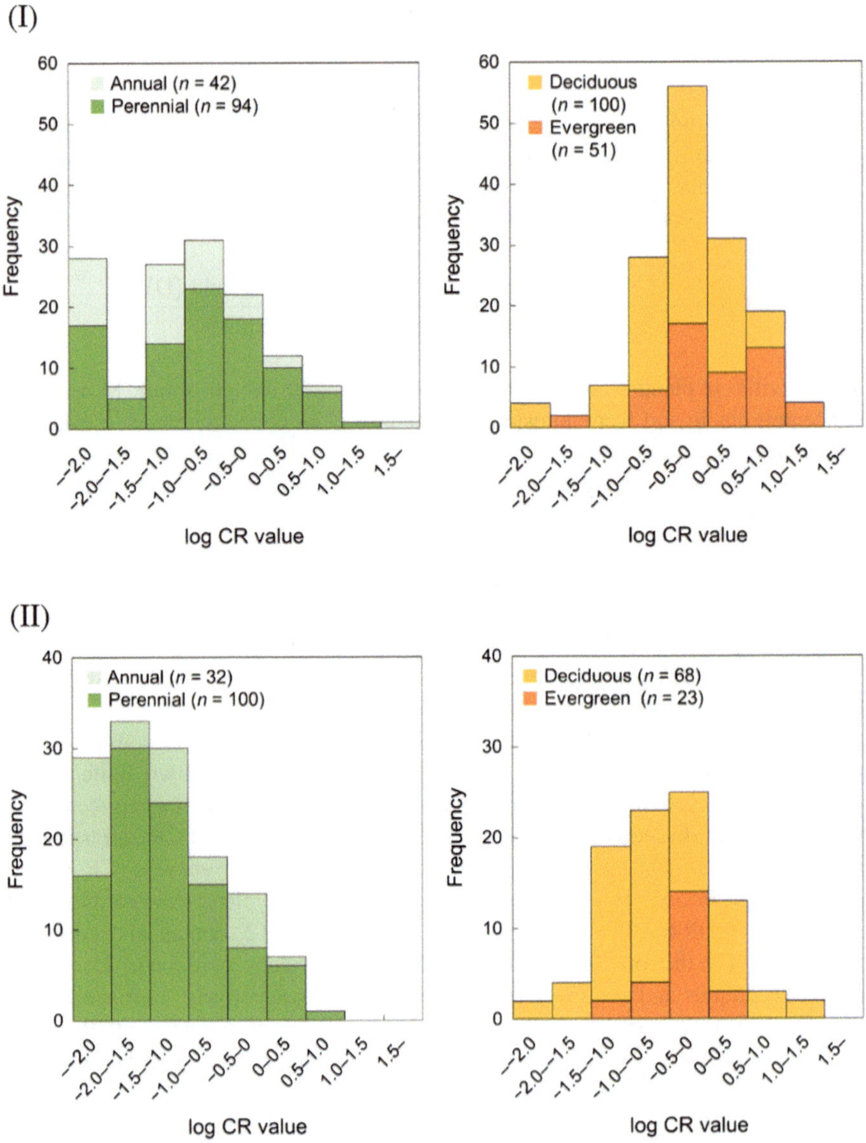

Fig. 3.4 Histograms of log CR values for samples collected in 2011 (I) and 2012 (II)

by Tagami et al. (2012) who reported that ^{137}Cs concentrations in new leaves decreased in the order evergreen trees > deciduous trees > herbaceous plants in 2011. In 2012, the number of evergreen trees showing high CR values decreased, whereas the occurrence of deciduous trees increased in the top 20 woody plants presenting high CR values (Table 3.3-II). Therefore, the surface absorption of ^{137}Cs through the leaf surface and its translocation to new organs at the initial stage of the

Table 3.3 Individuals presenting the top 20 CR values among herbaceous and woody plants collected in (I) 2011 and (II) 2012

I

Herbaceous plants

Sample name (scientific name)	Family	Life form	Sampling site	Rank of CR value		CR value			^{137}Cs concentration (Bq/kg DW)		
				Herbaceous samples	All samples	Value	±	Error[a]	Value	±	Error[b]
Chenopodium ficifolium	Chenopodiaceae	Annual	HIR1	1	1	54	±	4.4	4260	±	107
Houttuynia cordata	Saururaceae	Perennial	KOR4	2	6	10	±	1.6	12,700	±	260
Polygonum thunbergii	Polygonaceae	Annual	HIR1	3	16	4.7	±	0.69	370	±	46
Artemisia indica var. *maximowiczii*	Asteraceae	Perennial	SOM1	4	22	3.8	±	0.13	3050	±	52
Oenothera biennis	Onagraceae	Perennial	HIR1	5	23	3.6	±	0.50	287	±	33
Lilium auratum	Liliaceae	Perennial	KOR4	6	27	3.4	±	0.53	4150	±	150
Equisetum arvense	Equisetaceae	Perennial	SOM1	7	28	3.4	±	0.11	2700	±	32
Houttuynia cordata	Saururaceae	Perennial	KOR3	8	29	3.3	±	0.46	13,500	±	455
Achyranthes bidentata var. *tomentosa*	Amaranthaceae	Perennial	SOM1	9	30	3.3	±	0.54	2620	±	421
Houttuynia cordata	Saururaceae	Perennial	KOR4	10	35	2.9	±	0.44	3470	±	87
Houttuynia cordata	Saururaceae	Perennial	KOR4	11	38	2.5	±	0.38	2990	±	73
Pteridium aquilinum	Dennstaedtiaceae	Perennial	KOR3	12	40	2.4	±	0.38	9890	±	815
Artemisia indica var. *maximowiczii*	Asteraceae	Perennial	SOM2	13	42	2.4	±	0.55	276	±	23

(continued)

Table 3.3 (continued)

Perilla frutescens	Lamiaceae	Annual	SOM1	14	44	2.2	±	0.14	1700	±	98
Impatiens textori	Balsaminaceae	Perennial	SOM1	15	48	2.0	±	0.15	1600	±	106
Boehmeria silvestrii	Urticaceae	Perennial	DAT2	16	49	2.0	±	0.33	8460	±	166
Rumex japonicus	Polygonaceae	Perennial	DAT5	17	52	1.7	±	0.50	8890	±	991
Achillea millefolium	Asteraceae	Perennial	DAT5	18	55	1.7	±	0.45	8660	±	190
Trifolium repens	Fabaceae	Perennial	SOM1	19	58	1.5	±	0.16	1180	±	124
Bidens frondosa	Asteraceae	Annual	HIR1	20	60	1.3	±	0.43	107	±	33
Woody plants											
Sample name (scientific name)	Family	Life form	Sampling site	Rank of CR value		CR value			^{137}Cs concentration (Bq/kg DW)		
				Woody samples	All samples	Value	±	Error[a]	Value	±	Error[b]
Phyllostachys heterocycla	Poaceae	Evergreen	KOR3	1	2	30	±	4.0	120,000	±	1036
Eurya japonica	Theaceae	Evergreen	KOR5	2	3	26	±	5.1	49,000	±	753
Cryptomeria japonica	Taxodiaceae	Evergreen	SOM1	3	4	21	±	0.78	16,800	±	368
Euonymus japonicus	Celastraceae	Evergreen	KOR5	4	5	17	±	3.3	32,100	±	384
Camellia sasanqua	Theaceae	Evergreen	KOR5	5	7	8.9	±	1.7	16,600	±	273
Carpinus tschonoskii	Betulaceae	Deciduous	SOM1	6	8	8.7	±	1.4	6920	±	1057
Pieris japonica	Ericaceae	Evergreen	IWA2	7	9	8.4	±	1.4	18,000	±	333

Camellia sasanqua	Theaceae	Evergreen	KOR4	8	10	8.1	±	1.2	9820	±	219
Eurya japonica	Theaceae	Evergreen	IWA2	9	11	8.0	±	1.3	17,100	±	283
Abies firma	Pinaceae	Evergreen	SOM1	10	12	6.8	±	0.28	5360	±	150
Rubus palmatus var. *coptophyllus*	Rosaceae	Deciduous	SOM1	11	13	6.6	±	0.74	5210	±	564
Illicium anisatum	Illiciaceae	Evergreen	KOR5	12	14	4.7	±	0.92	8810	±	168
Callicarpa mollis	Verbenaceae	Deciduous	DAT1	13	15	4.7	±	1.2	23,500	±	231
Gardenia jasminoides	Rubiaceae	Evergreen	KOR5	14	17	4.3	±	0.84	8010	±	211
Sasamorpha borealis	Poaceae	Evergreen	SOM1	15	18	4.1	±	0.16	3240	±	80
Sasamorpha borealis	Poaceae	Evergreen	KAW1	16	19	4.0	±	0.77	9750	±	244
Pieris japonica	Ericaceae	Evergreen	KOR5	17	20	3.9	±	0.76	7310	±	191
Rhododendron sp.	Ericaceae	Evergreen	KAW5	18	21	3.8	±	0.56	10,300	±	1509
Orixa japonica	Rutaceae	Deciduous	SOM1	19	24	3.6	±	0.17	2840	±	109
Chamaecyparis obtusa	Cupressaceae	Evergreen	NAR1	20	25	3.6	±	0.65	10,700	±	202
II											
Herbaceous plants											
Sample name (scientific name)	Family	Life form	Sampling site	Rank of CR value		CR value			^{137}Cs concentration (Bq/kg DW)		
				Herbaceous samples	All samples	Value	±	Error[a]	Value	±	Error[b]
Oenothera biennis	Onagraceae	Perennial	NAR1	1	3	5.0	±	0.90	14,700	±	50

(continued)

Table 3.3 (continued)

Chenopodium ficifolium	Chenopodiaceae	Annual	HIR1	2	9	2.6	±	0.25	200	±	12
Pteridium aquilinum	Dennstaedtiaceae	Perennial	KOR3	3	11	2.3	±	0.30	8990	±	72
Solidago altissima	Asteraceae	Perennial	NAR1	4	14	2.2	±	0.39	6360	±	127
Boehmeria tricuspis	Urticaceae	Perennial	SOM1	5	17	2.0	±	0.07	1530	±	35
Houttuynia cordata	Saururaceae	Perennial	KOR4	6	19	1.3	±	0.20	1520	±	51
Houttuynia cordata	Saururaceae	Perennial	HIR1	7	21	1.2	±	0.21	90	±	14
Arisaema thunbergii spp. *urashima*	Araceae	Perennial	KOR4	8	23	1.1	±	0.18	1340	±	63
Lilium auratum	Liliaceae	Perennial	KOR4	9	31	0.93	±	0.14	1093	±	16
Chenopodium album	Chenopodiaceae	Annual	SOM2	10	35	0.80	±	0.18	92	±	5
Houttuynia cordata	Saururaceae	Perennial	KOR5	11	36	0.80	±	0.16	1449	±	32
Lamium album var. *barbatum*	Lamiaceae	Perennial	NAR1	12	37	0.80	±	0.15	2320	±	93
Osmunda japonica	Osmundaceae	Perennial	KOR4	13	38	0.73	±	0.11	866	±	11
Bidens frondosa	Asteraceae	Annual	HIR1	14	40	0.68	±	0.06	53	±	3
Achillea millefolium	Asteraceae	Perennial	DAT5	15	42	0.63	±	0.17	3132	±	24
Helianthus annuus	Asteraceae	Annual	IWA1	16	44	0.57	±	0.25	222	±	6
Boehmeria tricuspis	Urticaceae	Perennial	SOM1	17	48	0.48	±	0.02	369	±	8
Persicaria lapathifolia	Polygonaceae	Annual	HIR1	18	55	0.42	±	0.05	32	±	2

Sample name (scientific name)	Family	Life form	Sampling site	Rank of CR value		CR value			^{137}Cs concentration (Bq/kg DW)		
Lilium auratum	Liliaceae	Perennial	KOR5	19	56	0.41	±	0.08	747	±	33
Houttuynia cordata	Saururaceae	Perennial	NAR1	20	58	0.39	±	0.07	1135	±	54
Woody plants											
Sample name (scientific name)	Family	Life form	Sampling site	Rank of CR value		CR value			^{137}Cs concentration (Bq/kg DW)		
				Woody samples	All samples	Value	±	Error[a]	Value	±	Error[b]
Chengiopanax sciadophylloides	Araliaceae	Deciduous	KAW1	1	1	11	±	2.2	27,100	±	95
Acer crataegifolium	Aceraceae	Deciduous	KAW1	2	2	11	±	2.0	25,200	±	548
Acer crataegifolium	Aceraceae	Deciduous	KAW1	3	4	3.9	±	0.73	9140	±	127
Quercus aliena	Fagaceae	Deciduous	KAW1	4	5	3.6	±	0.68	8590	±	143
Chengiopanax sciadophylloides	Araliaceae	Deciduous	IWA2	5	6	3.3	±	0.55	6960	±	132
Ilex macropoda	Aquifoliaceae	Deciduous	KAW1	6	7	2.8	±	0.54	6730	±	42
Lithocarpus edulis	Fagaceae	Evergreen	IWA2	7	8	2.8	±	0.46	5810	±	133
Pinus thunbergii	Pinaceae	Evergreen	NAR1	8	10	2.5	±	0.45	7320	±	196
Chengiopanax sciadophylloides	Araliaceae	Deciduous	IWA2	9	12	2.3	±	0.37	4710	±	27
Viburnum furcatum	Caprifoliaceae	Deciduous	KAW1	10	13	2.2	±	0.42	5310	±	29
Wisteria floribunda	Fabaceae	Deciduous	SOM1	11	15	2.1	±	0.10	1640	±	59
Ilex macropoda	Aquifoliaceae	Deciduous	KAW1	12	16	2.1	±	0.39	4940	±	76

(continued)

Table 3.3 (continued)

Ulmus davidiana var. *japonica*	Ulmaceae	Deciduous	KAW1	13	18	1.8	±	0.35	4350	±	90
Fraxinus sieboldiana	Oleaceae	Deciduous	KAW1	14	20	1.3	±	0.24	2990	±	70
Alnus firma	Betulaceae	Deciduous	IWA2	15	22	1.2	±	0.19	2410	±	24
Eurya japonica	Theaceae	Evergreen	KOR4	16	24	1.1	±	0.17	1270	±	51
Viburnum furcatum	Caprifoliaceae	Deciduous	KAW1	17	25	1.1	±	0.20	2550	±	70
Acer crataegifolium	Aceraceae	Deciduous	IWA2	18	26	1.0	±	0.17	2110	±	18
Rhododendron sp.	Ericaceae	Evergreen	SOM1	19	27	0.98	±	0.03	765	±	11
Gamblea innovans	Araliaceae	Deciduous	IWA2	20	28	0.94	±	0.15	1950	±	26

[a]Associated uncertainty
[b]Measurement error

accident contributed to a great extent to its accumulation in evergreen trees. In addition, the ^{137}Cs concentration in some evergreen tree samples may have been overestimated because old and new leaves could not be differentiated. Most herbaceous plants in the top 20 individuals with high CR values were perennial herbs in 2011 and 2012. Several perennial herb species had leaves during the accident and would have been affected by ^{137}Cs surface absorption and translocation in 2011 (Hasegawa et al. 2009; Tagami et al. 2012). For instance, one individual of the perennial herb *Rumex japonicus* collected in 2011 showed much higher ^{137}Cs concentration (8890 Bq/kg DW) than other individuals (N.D.). It is inferred that this individual was affected by direct deposition. The rhizome of perennial herbs, such as *H. cordata*, may absorb bioavailable ^{137}Cs originating from the accident in 2011 and retain it even in the winter, resulting in higher ^{137}Cs concentrations in emerging perennial leaves in 2012 than in their annual equivalents.

Concentration ratios observed in 2012 decreased relative to 2011 (Fig. 3.4). In the case of the ChNPP, ^{137}Cs concentrations in plants decreased drastically in 1 year after the accident (Sawidis et al. 1990). Similar results were obtained after the FDNPP accident (Tagami et al. 2012; Mimura et al. 2014), consistent with the decrease in the amount of bioavailable ^{137}Cs in soil over time, as mentioned above. According to Takeda et al. (2013), Cs extractabilities by water and the ammonium acetate solution decreased with time, and the decreasing pattern was similar to that of TF. The high extraction rate recorded at site NAR1 would be one of the reasons for the high CR values of *O. biennis* and *Solidago altissima* (Table 3.3-II).

Figure 3.5 relates yearly changes in exchangeable ^{137}Cs and in CR. In general, yearly changes in extraction rate and in CR decreased in 2012 (Figs. 3.2 and 3.4). However, no apparent relationship was observed between these parameters ($r = -0.04$).

A linear relationship would be expected if CR values simply depended on exchangeable ^{137}Cs. However, no relationship was observed between these parameters, and CR values varied significantly, implying that CR values are influenced by other factors. Potential factors are species-specific capacities of ^{137}Cs accumulation, such as root growth rate and distribution, and the degree of mycorrhizal infection (Zhu and Smolders 2000; Vinichuk et al. 2013a, b). In addition, the concentration of exchangeable K affects CR (Zhu and Smolders 2000).

3.3.3 Cesium-137 Accumulator

In the present study, no clear ^{137}Cs accumulation was observed in Amaranthaceae, Chenopodiaceae, and Polygonaceae species, in contradiction with previous findings (Lasat et al. 1998; Broadley and Willey 1997; Broadley et al. 1999; Dushenkov et al. 1999; Fuhrmann et al. 2003). Previous results were obtained under conditions that differed from current settings, such as hydroponic and field experiments. Therefore, the investigated species may not exhibit high Cs accumulation in natural environments despite their ability to accumulate this metal. The *Chenopodium ficifolium*

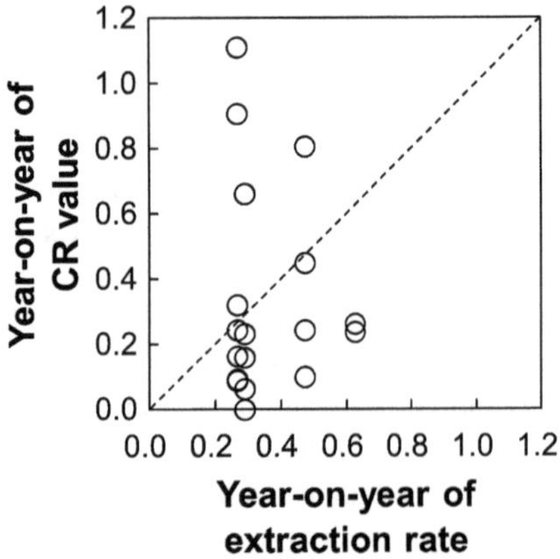

Fig. 3.5 Year-on-year (value in 2012/value in 2011) of extraction rate and CR value. Only CR values of herbaceous plants (13 species, 18 pairs) were used to eliminate the effect of surface absorption. Because no herbaceous plant was collected at site NAR1 in 2011, the data shown correspond to four sites (DAT2, KOR1, KOR2, and KOR3). The dashed line indicates the 1:1 line

samples showed high CR values collected at site HIR1, where ^{137}Cs concentrations in soil varied largely but no accumulation at site SOM2. Therefore, the accumulation ability of this species remains uncertain.

Evergreen trees, such as Theaceae species, displayed high CR values in 2011. Theaceae *C. sinensis* contaminated with ^{137}Cs derived from FDNPP was found in Shizuoka Prefecture in June 2011 (MHLW 2015). After detection, the ^{137}Cs concentration decreased drastically. It is considered that this decreases the results from the life form-dependence of the ^{137}Cs concentration, which is not species specific. Evergreen plants were affected by foliar absorption, and absorbed ^{137}Cs was translocated to the newly developed leaves.

Houttuynia cordata exhibited high CR values in 2011 and 2012 (Table 3.3), in agreement with a previous study by Yamashita et al. (2014). This result is of interest because several *H. cordata* individuals showed low CR values, and a survey by public institutions have found little ^{137}Cs accumulation in this species (MHLW 2015). In 2012, *C. sciadophylloides* showed high CR values at sites KAW1 and IWA2. Also, *C. sciadophylloides* shows high ^{137}Cs content in individuals collected in Nagano and Iwate Prefectures, which are far from FDNPP (MHLW 2015). Therefore, this species is expected to exhibit a species-specific ability to accumulate ^{137}Cs. *Acer crataegifolium* exhibited high CR values in both 2011 and 2012, suggesting that this species is also capable of species-specific accumulation.

Chengiopanax sciadophylloides is known as a manganese (Mn) accumulator and can accumulate Mn under a general soil condition (Memon et al. 1979; Mizuno et al. 2008). Although significant correlation between Mn and ^{137}Cs concentration in *C. sciadophylloides* was not found (Sugiura et al. 2016), endophytes in roots of *C. sciadophylloides* have an ability to increase bioavailability of Mn and ^{137}Cs uptake by *C. sciadophylloides* (Yamaji et al. 2016). Recent studies (Sugiura et al. 2016; Deguchi et al. 2017) indicated that the infection of endophytes in roots might have an important role for accumulation of ^{137}Cs in *C. sciadophylloides*. The identification of the species of endophyte contributing to ^{137}Cs accumulation and clarification of the mechanism should be necessary for the application of *C. sciadophylloides* to phytoremediation.

3.4 Conclusions

The CR values of various wild plants in Fukushima determined directly after the accident (2011) and the following year (2012) revealed that the CR values in 2012 declined as compared to that in 2011. In addition, woody plants presented higher CR values than herbaceous plants. This result is consistent with the absorption of ^{137}Cs through stem and/or leaf surfaces upon deposition and transfer to newly emerged tissues. Exchangeable ^{137}Cs rates in soils decreased at four sites because of the gradual stabilization of ^{137}Cs by clay minerals. Exchangeable ^{137}Cs concentrations and yearly changes differed between sampling sites, suggesting the influence of several factors, such as organic matter content, soil type, and mineralogical component. Both CR values and extraction rates decreased in 2012, implying that CR depends on changes in extraction rates as well as the initial effects of surface absorption of ^{137}Cs and plant characteristics, such as root distribution and the presence of mycorrhizal fungi. No clear ^{137}Cs accumulation ability was observed in Amaranthaceae, Chenopodiaceae, or Polygonaceae, which had been considered as ^{137}Cs accumulators. Some potential ^{137}Cs accumulators, such as *H. cordata*, *C. sciadophylloides*, and *A. crataegifolium*, were identified. Understanding the mechanism of ^{137}Cs accumulation will help to enable the use of these plants in phytoremediation.

3.5 Materials and Methods

3.5.1 Sample Collection

Surface soil samples (depth: 50 mm) in the plant habitats were collected using a core sampler (Ø 50 × H 50 mm) after removing bulky organic matters and stones from the surface.

Plant species were classified as herbaceous and woody plants. In turn, herbaceous plants were divided into annual and perennial herbs, while woody plants were divided into deciduous and evergreen trees. Bamboo species were categorized as evergreen trees. Newly emerged leaves were collected as measurement samples. Whenever possible, whole herbaceous plants were collected to ensure satisfactory masses for measurements. Current branches were collected as samples for woody plants, and only the leaves were used for ^{137}Cs concentration measurements. Samples obtained from May 2011 to January 2012 were defined as collected in 2011, and those obtained from March 2012 to November 2012 were defined as collected in 2012. Numbers of species and individuals are shown in Table 3.4.

Table 3.4 Numbers of collected species and individuals

Herbaceous/woody	Life form	2011		2012	
		Species	Individuals	Species	Individuals
Herbaceous	Annual	26	42	17	32
	Perennial	40	94	34	100
Woody	Deciduous	52	100	43	68
	Evergreen	29	51	14	23
	Total	147	287	108	223

3.5.2 Sample Preparation

Soil samples were air-dried for a few weeks and subsequently passed through a 2-mm sieve. Samples for exchangeable ^{137}Cs measurement were shaken with 1 M ammonium acetate (1:10) for 2 h to extract ^{137}Cs. The suspensions were centrifuged at 3500 rpm for 10 min, and the supernatant were filtrated by a 0.45-μm membrane filter.

Plant samples were washed with running water to remove soil particles and subsequently rinsed with distilled water. Samples for autoradiographic analysis were dried with paper towels and packed in plastic bags. Samples for ^{137}Cs concentration measurement were dried in paper bags at 80 °C for 48 h and ground with a mill mixer.

3.5.3 Sample Analysis

Soil and extraction samples were enclosed in plastic containers (Ø 56 × H 64 mm; U-8 container PP-U8, Sekiya Rika), and ^{137}Cs activities were measured using a high-purity germanium (HPGe) detector (GC3520, Canberra; GEM-30195-P, Seiko EG&G Ortec).

Autoradiograph images were obtained by placing imaging plates (BAS-IP MS 2025 E, Fujifilm) on the packed plant samples for 48–90 h and developing it (Typhoon FLA 9000, GE Healthcare). Powdered plant samples for ^{137}Cs concentration measurement were enclosed in plastic containers (Ø 56 × H 35 mm; U-9 container PP-U9, Sekiya Rika) or round-bottomed tubes (Ø 13 × H 72 mm; JB2000, Eiken Chemical) depending on their mass (>1.0 g: U-9 container; <1.0 g: round-bottomed tube). Next, ^{137}Cs activities were measured with an HPGe detector (U-9 container: GC3520 or GEM-30195-P; round-bottomed tube: GWL-300-15-S, Seiko EG&G Ortec). Several samples (>1.0 g) were enclosed into polyethylene vial bottles (Ø 27 × H 60 mm; Polyvials V-natural high density polyethylene, Zinsser Analytic), and ^{137}Cs activities were measured using a thallium-doped sodium iodide [NaI(TI)] scintillation detector (ARC-7001, Hitachi Aloka Medical). All of the analytical data for plant samples collected in 2011 and 2012 are attached in Appendices.

3.5.4 Calculation and Statistical Treatment

Spectra recorded using HPGe and NaI(Tl) scintillation detectors were analyzed and calculated using Gamma Studio Version 2.1.1. (Canberra) and ARC-7001 software, respectively. Continuous ^{137}Cs measurements were performed until the time exceeded 18,000 s or the error count of ^{137}Cs was within 3% of the net count. Samples evaluated as less than the detection limit were written as "not detected (N.D.)" and handled as 0 Bq in calculations. The ^{137}Cs activities in soil were corrected for decay to April 1, 2011, and their means at that same site were calculated afterward. Because no ^{137}Cs concentration difference was observed between 2011 and 2012 (t test, $p < 0.05$), the results for both years were combined to increase the number of replicates ($n = 3$–20). The difference in the ^{137}Cs activity for decay from May 2011 to October 2012 approximated 3%, which was sufficiently lower than the dispersion between replicates. The Smirnov-Grubbs test for rejection was performed ($p < 0.05$) before calculating means.

The yearly change in CR was calculated as the ratio between CR values found in 2012 and 2011 for each plant species collected from the same site. Only the CR values of the herbaceous plants were used to exclude effect of direct deposition. In the case of multiple collections of the same species, the means were used. Correlations between yearly changes in extraction rate and CR value were assessed using Spearman's rank correlation coefficient ($p < 0.05$).

Appendices

Appendix 1: Cesium-137 Concentration of Plant Leaves and Its Concentration Ratio (CR) in 2011

Sample name (Scientific name)	Family	Herbaceous/ woody	Lifeform	Sampling site	^{137}Cs concentration (Bq/kg DW)			CR value		
					Value	±	Error[a]	Value	±	Error[b]
(Unknoun)	Fabaceae	Woody	Deciduous	DAT5	12,300	±	1300	2.41	±	0.68
(Unknoun)	Poaceae	Woody	Evergreen	DAT3	4700	±	235	0.86	±	0.14
(Unknoun)	Poaceae	Woody	Evergreen	SOM1	533	±	44	0.68	±	0.06
(Unknoun)	Fabaceae	Woody	Deciduous	HIR2	3930	±	619	0.44	±	0.07
(Unknoun)	Brassicaceae	Herbaceous	Annual	KOR3	1090	±	106	0.27	±	0.04
(Unknoun)	Poaceae	Herbaceous	Perennial	KOR3	527	±	93	0.13	±	0.03
(Unknoun)	Poaceae	Woody	Evergreen	DAT1	564	±	81	0.11	±	0.03
(Unknoun)	Poaceae	Woody	Evergreen	FUT1	1350	±	61	0.02	±	0.00
(Unknoun)	Asteraceae	Herbaceous	Perennial	KOR1	N.D.	±	–	0.00	±	–
Abies firma	Pinaceae	Woody	Evergreen	SOM1	5360	±	150	6.76	±	0.28
Acer crataegifolium	Aceraceae	Woody	Deciduous	KAW1	8330	±	631	3.43	±	0.70
Acer crataegifolium	Aceraceae	Woody	Deciduous	KAW5	3140	±	45	1.17	±	0.02
Acer palmatum	Aceraceae	Woody	Deciduous	KAW6	2240	±	412	0.50	±	0.09

Acer palmatum	Aceraceae	Woody	Deciduous	KAW6	1800	±	34	0.40	±	0.01
Acer rufinerve	Aceraceae	Woody	Deciduous	DAT3	7090	±	210	1.29	±	0.21
Acer rufinerve	Aceraceae	Woody	Deciduous	DAT3	5960	±	132	1.09	±	0.17
Acer shirasawanum	Aceraceae	Woody	Deciduous	KOR4	2750	±	93	2.28	±	0.35
Achillea millefolium	Asteraceae	Herbaceous	Perennial	DAT5	8660	±	190	1.71	±	0.45
Achillea millefolium	Asteraceae	Herbaceous	Perennial	DAT5	4980	±	121	0.98	±	0.26
Achyranthes bidentata var. *tomentosa*	Amaranthaceae	Herbaceous	Perennial	SOM1	2620	±	421	3.31	±	0.54
Achyranthes bidentata var. *tomentosa*	Amaranthaceae	Herbaceous	Perennial	DAT5	560	±	56	0.11	±	0.03
Achyranthes bidentata var. *tomentosa*	Amaranthaceae	Herbaceous	Perennial	KAW2	478	±	41	0.07	±	0.01
Achyranthes bidentata var. *tomentosa*	Amaranthaceae	Herbaceous	Perennial	KOR2	320	±	47	0.05	±	0.02
Actinidia deliciosa	Actinidiaceae	Woody	Deciduous	FUT1	3820	±	84	0.05	±	0.01
Adenophora triphylla var. *japonica*	Campanulaceae	Herbaceous	Perennial	KAW4	N.D.	±	–	0.00	±	–
Akebia quinata	Lardizabalaceae	Woody	Deciduous	KOR2	778	±	45	0.13	±	0.04
Akebia quinata	Lardizabalaceae	Woody	Deciduous	SOM2	N.D.	±	–	0.00	±	–
Alnus firma	Betulaceae	Woody	Deciduous	IWA2	2220	±	127	1.04	±	0.18
Alnus firma	Betulaceae	Woody	Deciduous	NAR1	1050	±	45	0.35	±	0.07
Amaranthus lividus	Amaranthaceae	Herbaceous	Annual	KAW2	2850	±	106	0.40	±	0.06
Amaranthus lividus	Amaranthaceae	Herbaceous	Annual	KAW3	810	±	63	0.05	±	0.02
Amaranthus patulus	Amaranthaceae	Herbaceous	Annual	KOR1	930	±	46	0.15	±	0.02

(continued)

Sample name (Scientific name)	Family	Herbaceous/ woody	Lifeform	Sampling site	^{137}Cs concentration (Bq/kg DW)			CR value		
					Value	±	Error[a]	Value	±	Error[b]
Amaranthus patulus	Amaranthaceae	Herbaceous	Annual	DAT5	297	±	39	0.06	±	0.02
Amaranthus patulus	Amaranthaceae	Herbaceous	Annual	KAW3	840	±	49	0.05	±	0.02
Amaranthus patulus	Amaranthaceae	Herbaceous	Annual	KAW2	298	±	58	0.04	±	0.01
Amaranthus patulus	Amaranthaceae	Herbaceous	Annual	KAW2	N.D.	±	–	0.00	±	–
Ambrosia trifida	Asteraceae	Herbaceous	Annual	KOR1	350	±	46	0.06	±	0.01
Ambrosia trifida	Asteraceae	Herbaceous	Annual	KOR2	137	±	50	0.02	±	0.01
Ambrosia trifida	Asteraceae	Herbaceous	Annual	KOR2	N.D.	±	–	0.00	±	–
Ampelopsis glandulosa var. *heterophylla*	Vitaceae	Woody	Deciduous	SOM1	627	±	74	0.79	±	0.10
Aralia elata	Araliaceae	Woody	Deciduous	KOR3	989	±	308	0.24	±	0.08
Aralia elata	Araliaceae	Woody	Deciduous	DAT3	415	±	52	0.08	±	0.02
Artemisia indica var. *moximowiczii*	Asteraceae	Herbaceous	Perennial	SOM1	3050	±	52	3.84	±	0.13
Artemisia indica var. *moximowiczii*	Asteraceae	Herbaceous	Perennial	SOM2	276	±	23	2.35	±	0.55
Artemisia indica var. *moximowiczii*	Asteraceae	Herbaceous	Perennial	SOM1	326	±	11	0.41	±	0.02
Artemisia indica var. *moximowiczii*	Asteraceae	Herbaceous	Perennial	KOR3	1550	±	34	0.38	±	0.05
Artemisia indica var. *moximowiczii*	Asteraceae	Herbaceous	Perennial	KOR1	190	±	8	0.03	±	0.00
Aster yomena	Asteraceae	Herbaceous	Perennial	KAW4	1480	±	632	0.46	±	0.20

Aucuba japonica	Cornaceae	Woody	Evergreen	KAW1	4450	±	547	1.83	±	0.41
Aucuba japonica	Cornaceae	Woody	Evergreen	KOR3	2130	±	63	0.54	±	0.07
Benthamidia florida	Cornaceae	Woody	Deciduous	DAT2	730	±	55	0.17	±	0.03
Bidens frondosa	Asteraceae	Herbaceous	Annual	HIR1	107	±	33	1.35	±	0.43
Boehmeria silvestrii	Urticaceae	Herbaceous	Perennial	DAT2	8460	±	166	1.95	±	0.33
Boehmeria silvestrii	Urticaceae	Herbaceous	Perennial	DAT2	3410	±	559	0.79	±	0.18
Boehmeria tricuspis	Urticaceae	Herbaceous	Perennial	SOM1	632	±	66	0.79	±	0.09
Boehmeria tricuspis	Urticaceae	Herbaceous	Perennial	DAT2	2220	±	510	0.51	±	0.15
Boehmeria tricuspis	Urticaceae	Herbaceous	Perennial	KOR3	2020	±	83	0.50	±	0.07
Broussonetia kazinoki	Moraceae	Woody	Deciduous	SOM1	2190	±	174	2.75	±	0.23
Broussonetia kazinoki	Moraceae	Woody	Deciduous	SOM1	219	±	49	0.28	±	0.06
Callicarpa mollis	Verbenaceae	Woody	Deciduous	DAT1	23,500	±	231	4.72	±	1.25
Callicarpa mollis	Verbenaceae	Woody	Deciduous	DAT1	1910	±	60	0.39	±	0.10
Camellia japonica	Theaceae	Woody	Evergreen	KOR4	3750	±	106	3.10	±	0.47
Camellia japonica	Theaceae	Woody	Evergreen	NAR1	2880	±	53	0.97	±	0.17
Camellia sasanqua	Theaceae	Woody	Evergreen	KOR5	16,600	±	273	8.91	±	1.72
Camellia sasanqua	Theaceae	Woody	Evergreen	KOR4	9820	±	219	8.12	±	1.23
Camellia sasanqua	Theaceae	Woody	Evergreen	FUT1	25,900	±	272	0.34	±	0.07
Camellia sinensis	Theaceae	Woody	Evergreen	KAW2	14,500	±	305	2.04	±	0.28

(continued)

Sample name (Scientific name)	Family	Herbaceous/ woody	Lifeform	Sampling site	^{137}Cs concentration (Bq/kg DW)			CR value		
					Value	±	Error[a]	Value	±	Error[b]
Camellia sinensis	Theaceae	Woody	Evergreen	KAW2	7710	±	235	1.09	±	0.15
Campanula punctata	Campanulaceae	Herbaceous	Perennial	KOR3	2940	±	96	0.73	±	0.10
Carex sp.	Cyperaceae	Herbaceous	Perennial	DAT1	3030	±	132	0.61	±	0.16
Carpinus tschonoskii	Betulaceae	Woody	Deciduous	SOM1	6920	±	1057	8.70	±	1.35
Carpinus tschonoskii	Betulaceae	Woody	Deciduous	SOM1	1350	±	76	1.70	±	0.11
Castanea crenata	Fagaceae	Woody	Deciduous	DAT5	12,700	±	251	2.49	±	0.65
Castanea crenata	Fagaceae	Woody	Deciduous	DAT1	2830	±	34	0.57	±	0.15
Castanea crenata	Fagaceae	Woody	Deciduous	DAT1	1110	±	42	0.22	±	0.06
Cerasus jamasakura	Rosaceae	Woody	Deciduous	DAT5	4900	±	176	0.96	±	0.25
Cerasus jamasakura	Rosaceae	Woody	Deciduous	DAT2	435	±	11	0.10	±	0.02
Cerasus speciosa	Rosaceae	Woody	Deciduous	KOR3	752	±	57	0.19	±	0.03
Cerasus speciosa	Rosaceae	Woody	Deciduous	KOR3	472	±	36	0.12	±	0.02
Cerasus yedoensis	Rosaceae	Woody	Deciduous	KAW6	1,990	±	507	0.44	±	0.11
Cerasus yedoensis	Rosaceae	Woody	Deciduous	HIR2	2590	±	122	0.29	±	0.01
Cerasus yedoensis	Rosaceae	Woody	Deciduous	KAW6	1150	±	287	0.25	±	0.06
Cerasus yedoensis	Rosaceae	Woody	Deciduous	HIR2	1220	±	21	0.14	±	0.00
Chamaecyparis obtusa	Cupressaceae	Woody	Evergreen	NAR1	10,700	±	202	3.59	±	0.65

Chelidonium majus var. *asiaticum*	Papaveraceae	Herbaceous	Perennial	KOR2	1010	±	126	0.17	±	0.05
Chengiopanax sciadophylloides	Araliaceae	Woody	Deciduous	KAW5	1610	±	36	0.60	±	0.01
Chengiopanax sciadophylloides	Araliaceae	Woody	Deciduous	KAW4	1570	±	28	0.49	±	0.01
Chenopodium album	Chenopodiaceae	Herbaceous	Annual	HIR2	816	±	78	0.09	±	0.01
Chenopodium album	Chenopodiaceae	Herbaceous	Annual	HIR2	550	±	52	0.06	±	0.01
Chenopodium album	Chenopodiaceae	Herbaceous	Annual	KAW2	68	±	26	0.01	±	0.00
Chenopodium album	Chenopodiaceae	Herbaceous	Annual	HIR1	N.D.	±	–	0.00	±	–
Chenopodium album	Chenopodiaceae	Herbaceous	Annual	SOM2	N.D.	±	–	0.00	±	–
Chenopodium album	Chenopodiaceae	Herbaceous	Annual	DAT2	N.D.	±	–	0.00	±	–
Chenopodium ambrosioides	Chenopodiaceae	Herbaceous	Annual	HIR1	N.D.	±	–	0.00	±	–
Chenopodium ficifolium	Chenopodiaceae	Herbaceous	Annual	HIR1	4260	±	107	53.6	±	4.38
Cirsium japonicum	Asteraceae	Herbaceous	Perennial	KAW4	1170	±	521	0.36	±	0.16
Cirsium sp.	Asteraceae	Herbaceous	Perennial	DAT2	388	±	54	0.09	±	0.02
Citrus junos	Rutaceae	Woody	Evergreen	FUT1	8670	±	145	0.11	±	0.02
Clethra barbinervis	Clethraceae	Woody	Deciduous	KAW5	3750	±	70	1.39	±	0.03
Clethra barbinervis	Clethraceae	Woody	Deciduous	NAR1	1460	±	68	0.49	±	0.09
Commelina sp.	Commelinaceae	Herbaceous	Annual	HIR2	483	±	24	0.05	±	0.00
Commelina sp.	Commelinaceae	Herbaceous	Annual	HIR2	395	±	134	0.04	±	0.02
Conyza bonariensis	Asteraceae	Herbaceous	Annual	HIR2	404	±	12	0.05	±	0.00

(continued)

Sample name (Scientific name)	Family	Herbaceous/ woody	Lifeform	Sampling site	^{137}Cs concentration (Bq/kg DW)			CR value		
					Value	±	Error[a]	Value	±	Error[b]
Cosmos sulphureus	Asteraceae	Herbaceous	Annual	DAT5	298	±	43	0.06	±	0.02
Crypromeria japonica	Taxodiaceae	Woody	Evergreen	SOM1	16,800	±	368	21.1	±	0.78
Crypromeria japonica	Taxodiaceae	Woody	Evergreen	KOR3	4260	±	432	1.06	±	0.18
Crypromeria japonica	Taxodiaceae	Woody	Evergreen	KOR3	3130	±	63	0.78	±	0.11
Crypromeria japonica	Taxodiaceae	Woody	Evergreen	KOR3	2460	±	390	0.61	±	0.13
Crypromeria japonica	Taxodiaceae	Woody	Evergreen	KOR3	2150	±	65	0.54	±	0.07
Crypromeria japonica	Taxodiaceae	Woody	Evergreen	KOR3	1920	±	38	0.48	±	0.06
Crypromeria japonica	Taxodiaceae	Woody	Evergreen	KOR3	1110	±	66	0.28	±	0.04
Diospyros kaki	Ebenaceae	Woody	Deciduous	FUT1	6000	±	158	0.08	±	0.02
Enkianthus perulatus	Ericaceae	Woody	Deciduous	HIR2	2250	±	37	0.25	±	0.00
Equisetum arvense	Equisetaceae	Herbaceous	Perennial	SOM1	2700	±	32	3.40	±	0.11
Equisetum arvense	Equisetaceae	Herbaceous	Perennial	KOR1	56	±	7	0.01	±	0.00
Equisetum arvense	Equisetaceae	Herbaceous	Perennial	SOM2	N.D.	±	–	0.00	±	–
Equisetum arvense	Equisetaceae	Herbaceous	Perennial	KOR3	N.D.	±	–	0.00	±	–
Euonymus japonicus	Celastraceae	Woody	Evergreen	KOR5	32,100	±	384	17.3	±	3.33
Eurya japonica	Theaceae	Woody	Evergreen	KOR5	49,000	±	753	26.3	±	5.09
Eurya japonica	Theaceae	Woody	Evergreen	IWA2	17,100	±	283	7.99	±	1.31

Fagopyrum esculentum	Polygonaceae	Herbaceous	Annual	KOR1	N.D.	±	–	0.00	±	–
Farfugium japonicum	Asteraceae	Herbaceous	Perennial	FUT1	9230	±	210	0.12	±	0.03
Fraxinus sieboldiana	Oleaceae	Woody	Deciduous	KAW5	1970	±	35	0.73	±	0.01
Fraxinus sieboldiana	Oleaceae	Woody	Deciduous	KAW4	1020	±	362	0.32	±	0.11
Gardenia jasminoides	Rubiaceae	Woody	Evergreen	KOR5	8010	±	211	4.32	±	0.84
Hamamlis japonica	Hamamelidaceae	Woody	Deciduous	DAT3	3260	±	60	0.60	±	0.10
Hamamlis japonica	Hamamelidaceae	Woody	Deciduous	DAT3	2170	±	290	0.40	±	0.08
Helianthus annuus	Asteraceae	Herbaceous	Annual	DAT5	1040	±	44	0.21	±	0.05
Helianthus annuus	Asteraceae	Herbaceous	Annual	KOR1	367	±	37	0.06	±	0.01
Helianthus tuberosus	Asteraceae	Herbaceous	Perennial	DAT1	1000	±	286	0.20	±	0.08
Helianthus tuberosus	Asteraceae	Herbaceous	Perennial	DAT1	875	±	46	0.18	±	0.05
Helianthus tuberosus	Asteraceae	Herbaceous	Perennial	KOR2	542	±	36	0.09	±	0.03
Houttuynia cordata	Saururaceae	Herbaceous	Perennial	KOR4	12,700	±	260	10.5	±	1.59
Houttuynia cordata	Saururaceae	Herbaceous	Perennial	KOR3	13,500	±	455	3.33	±	0.46
Houttuynia cordata	Saururaceae	Herbaceous	Perennial	KOR4	3470	±	87	2.88	±	0.44
Houttuynia cordata	Saururaceae	Herbaceous	Perennial	KOR4	2990	±	73	2.48	±	0.38
Houttuynia cordata	Saururaceae	Herbaceous	Perennial	KOR3	1960	±	78	0.49	±	0.07
Houttuynia cordata	Saururaceae	Herbaceous	Perennial	DAT4	716	±	47	0.18	±	0.02
Houttuynia cordata	Saururaceae	Herbaceous	Perennial	KOR3	437	±	59	0.11	±	0.02

(continued)

Sample name (Scientific name)	Family	Herbaceous/ woody	Lifeform	Sampling site	^{137}Cs concentration (Bq/kg DW)			CR value		
					Value	±	Error[a]	Value	±	Error[b]
Houttuynia cordata	Saururaceae	Herbaceous	Perennial	KOR3	240	±	38	0.06	±	0.01
Houttuynia cordata	Saururaceae	Herbaceous	Perennial	KOR3	236	±	42	0.06	±	0.01
Humulus japonicus	Moraceae	Herbaceous	Annual	KOR2	650	±	53	0.11	±	0.03
Hydrangea macrophylla	Saxifragaceae	Woody	Deciduous	DAT2	1470	±	23	0.34	±	0.06
Hydrangea macrophylla	Saxifragaceae	Woody	Deciduous	SOM1	109	±	24	0.14	±	0.03
Hydrangea macrophylla	Saxifragaceae	Woody	Deciduous	KAW6	N.D.	±	–	0.00	±	–
Illicium anisatum	Illiciaceae	Woody	Evergreen	KOR5	8810	±	168	4.74	±	0.92
Illicium anisatum	Illiciaceae	Woody	Evergreen	KOR5	1250	±	57	0.68	±	0.13
Impatiens textori	Balsaminaceae	Herbaceous	Perennial	SOM1	1600	±	106	2.03	±	0.15
Indigofera pseudotinctoria	Fabaceae	Woody	Deciduous	NAR1	1520	±	54	0.51	±	0.09
Juglans mandshurica var. *sachalinensis*	Juglandaceae	Woody	Deciduous	KOR3	755	±	84	0.19	±	0.03
Juglans mandshurica var. *sachalinensis*	Juglandaceae	Woody	Deciduous	DAT3	726	±	52	0.13	±	0.02
Lactuca indica	Asteraceae	Herbaceous	Annual	DAT1	1130	±	23	0.23	±	0.06
Larix kaempferi	Pinaceae	Woody	Deciduous	DAT5	3860	±	446	0.76	±	0.22
Lespedeza bicolor	Fabaceae	Woody	Deciduous	DAT1	1890	±	388	0.38	±	0.13
Lespedeza bicolor	Fabaceae	Woody	Deciduous	DAT1	1120	±	37	0.23	±	0.06

Lespedeza crytobotrya	Fabaceae	Woody	Deciduous	KAW5	3120	±	967	1.16	±	0.36
Ligustrum lucidum	Oleaceae	Woody	Evergreen	KOR5	1300	±	58	0.71	±	0.14
Lilium auratum	Liliaceae	Herbaceous	Perennial	KOR4	4150	±	150	3.43	±	0.53
Lindera umbellata	Lauraceae	Woody	Deciduous	DAT3	5920	±	290	1.08	±	0.18
Lysimachia clethroides	Primulaceae	Herbaceous	Perennial	KOR3	1790	±	132	0.44	±	0.07
Lysimachia clethroides	Primulaceae	Herbaceous	Perennial	DAT1	1460	±	325	0.29	±	0.10
Macleaya cordata	Papaveraceae	Herbaceous	Perennial	DAT5	590	±	74	0.12	±	0.03
Magnolia kobus	Magnoliaceae	Woody	Deciduous	KOR3	2680	±	102	0.66	±	0.09
Mahonia japonica	Berberidaceae	Woody	Evergreen	FUK1	1790	±	52	1.06	±	0.39
Mallotus japonicus	Euphorbiaceae	Woody	Deciduous	IWA2	3800	±	121	1.78	±	0.30
Morus bombycis	Moraceae	Woody	Deciduous	DAT1	596	±	22	0.12	±	0.03
Oenothera biennis	Onagraceae	Herbaceous	Perennial	HIR1	287	±	33	3.61	±	0.50
Oenothera biennis	Onagraceae	Herbaceous	Perennial	KOR1	120	±	31	0.02	±	0.01
Orixa japonica	Rutaceae	Woody	Deciduous	SOM1	2840	±	109	3.59	±	0.17
Oryza sativa	Poaceae	Herbaceous	Annual	DAT4	425	±	66	0.11	±	0.02
Osmanthus fragrans var. *aurantiacus*	Oleaceae	Woody	Evergreen	NAR1	1370	±	50	0.46	±	0.08
Osmunda japonica	Osmundaceae	Herbaceous	Perennial	KOR3	1030	±	86	0.26	±	0.04
Perilla frutescens	Lamiaceae	Herbaceous	Annual	SOM1	1700	±	98	2.15	±	0.14
Persicaria lapathifolia	Polygonaceae	Herbaceous	Annual	HIR1	N.D.	±	–	0.00	±	–

(continued)

Sample name (Scientific name)	Family	Herbaceous/ woody	Lifeform	Sampling site	^{137}Cs concentration (Bq/kg DW)			CR value		
					Value	±	Error[a]	Value	±	Error[b]
Persicaria nipponensis	Polygonaceae	Herbaceous	Annual	KOR3	1860	±	148	0.46	±	0.07
Persicaria senticosa	Polygonaceae	Herbaceous	Annual	KOR2	2070	±	912	0.35	±	0.19
Petasites japonicus	Asteraceae	Herbaceous	Perennial	FUK1	366	±	22	0.22	±	0.08
Petasites japonicus	Asteraceae	Herbaceous	Perennial	FUK1	349	±	9	0.21	±	0.08
Petasites japonicus	Asteraceae	Herbaceous	Perennial	KOR3	837	±	58	0.21	±	0.03
Petasites japonicus	Asteraceae	Herbaceous	Perennial	DAT2	757	±	23	0.17	±	0.03
Petasites japonicus	Asteraceae	Herbaceous	Perennial	DAT2	743	±	38	0.17	±	0.03
Petunia sp.	Solanaceae	Herbaceous	Annual	DAT5	602	±	69	0.12	±	0.03
Photinia glabra	Rosaceae	Woody	Evergreen	NAR1	921	±	42	0.31	±	0.06
Phragmites japonica	Poaceae	Herbaceous	Perennial	KOR2	35	±	3	0.01	±	0.00
Phyllostachys heterocycla	Poaceae	Woody	Evergreen	KOR3	120,000	±	1036	29.6	±	3.96
Phyllostachys heterocycla	Poaceae	Woody	Evergreen	KOR3	1150	±	83	0.29	±	0.04
Pieris japonica	Ericaceae	Woody	Evergreen	IWA2	18,000	±	333	8.43	±	1.39
Pieris japonica	Ericaceae	Woody	Evergreen	KOR5	7310	±	191	3.93	±	0.76
Pinus densifolia	Pinaceae	Woody	Evergreen	NAR1	2820	±	83	0.95	±	0.17
Polygonatum falcatum	Liliaceae	Herbaceous	Perennial	KAW5	2910	±	160	1.08	±	0.06
Polygonatum falcatum	Liliaceae	Herbaceous	Perennial	KAW6	N.D.	±	–	0.00	±	–

Polygonum sp.	Polygonaceae	Herbaceous	Annual	DAT2	2850	±	149	0.66	±	0.11
Polygonum sp.	Polygonaceae	Herbaceous	Annual	DAT2	558	±	58	0.13	±	0.03
Polygonum sp.	Polygonaceae	Herbaceous	Annual	KAW3	359	±	11	0.02	±	0.01
Polygonum thunbergii	Polygonaceae	Herbaceous	Annual	HIR1	370	±	46	4.65	±	0.69
Populus tremula var. *sieboldii*	Salicaceae	Woody	Deciduous	KOR3	6990	±	172	1.73	±	0.23
Pteridium aquilinum	Dennstaedtiaceae	Herbaceous	Perennial	KOR3	9890	±	815	2.45	±	0.38
Pteridium aquilinum	Dennstaedtiaceae	Herbaceous	Perennial	DAT5	1210	±	400	0.24	±	0.10
Pteridium aquilinum	Dennstaedtiaceae	Herbaceous	Perennial	DAT2	149	±	12	0.03	±	0.01
Pteridium aquilinum	Dennstaedtiaceae	Herbaceous	Perennial	DAT2	108	±	8	0.02	±	0.00
Pteridium aquilinum	Dennstaedtiaceae	Herbaceous	Perennial	KAW6	N.D.	±	–	0.00	±	–
Pueraria lobata	Fabaceae	Herbaceous	Perennial	KOR3	2260	±	662	0.56	±	0.18
Pueraria lobata	Fabaceae	Herbaceous	Perennial	KOR2	1230	±	80	0.21	±	0.06
Pueraria lobata	Fabaceae	Herbaceous	Perennial	SOM1	N.D.	±	–	0.00	±	–
Pueraria lobata	Fabaceae	Herbaceous	Perennial	DAT1	N.D.	±	–	0.00	±	–
Pueraria lobata	Fabaceae	Herbaceous	Perennial	KOR2	N.D.	±	–	0.00	±	–
Quercus aliena	Fagaceae	Woody	Deciduous	KAW1	7810	±	659	3.22	±	0.66
Quercus aliena	Fagaceae	Woody	Deciduous	KAW1	4950	±	501	2.04	±	0.44
Quercus crispula	Fagaceae	Woody	Deciduous	DAT3	4,210	±	126	0.77	±	0.12
Quercus crispula	Fagaceae	Woody	Deciduous	DAT3	2840	±	389	0.52	±	0.11

(continued)

Sample name (Scientific name)	Family	Herbaceous/ woody	Lifeform	Sampling site	^{137}Cs concentration (Bq/kg DW)			CR value		
					Value	±	Error[a]	Value	±	Error[b]
Quercus myrsinifolia	Fagaceae	Woody	Evergreen	FUK1	2620	±	98	1.56	±	0.57
Quercus phillyraeoides	Fagaceae	Woody	Evergreen	FUK1	893	±	52	0.53	±	0.20
Quercus serrata	Fagaceae	Woody	Deciduous	DAT1	4810	±	764	0.97	±	0.30
Quercus serrata	Fagaceae	Woody	Deciduous	KAW4	2800	±	568	0.87	±	0.18
Quercus serrata	Fagaceae	Woody	Deciduous	SOM1	485	±	55	0.61	±	0.07
Quercus serrata	Fagaceae	Woody	Deciduous	DAT3	3310	±	421	0.60	±	0.12
Quercus serrata	Fagaceae	Woody	Deciduous	DAT3	2860	±	48	0.52	±	0.08
Quercus serrata	Fagaceae	Woody	Deciduous	KOR3	1470	±	117	0.36	±	0.06
Quercus serrata	Fagaceae	Woody	Deciduous	KAW6	1140	±	34	0.25	±	0.01
Quercus serrata	Fagaceae	Woody	Deciduous	KAW6	N.D.	±	–	0.00	±	–
Quercus sp.	Fagaceae	Woody	Deciduous	KAW5	3140	±	107	1.17	±	0.04
Ranunculus cantoniensis	Ranunculaceae	Herbaceous	Perennial	KOR3	1340	±	76	0.33	±	0.05
Ranunculus cantoniensis	Ranunculaceae	Herbaceous	Perennial	KOR1	606	±	76	0.10	±	0.02
Reynoutria japonica	Polygonaceae	Herbaceous	Perennial	DAT2	746	±	245	0.17	±	0.06
Reynoutria japonica	Polygonaceae	Herbaceous	Perennial	SOM1	N.D.	±	–	0.00	±	–
Reynoutria japonica	Polygonaceae	Herbaceous	Perennial	DAT5	N.D.	±	–	0.00	±	–
Rhaphiolepis ndica var. *umbellata*	Rosaceae	Woody	Evergreen	FUT1	2070	±	56	0.03	±	0.01

Rhododendron × pulchrum	Ericaceae	Woody	Evergreen	NAR1	3660	±	92	1.23	±	0.22
Rhododendron × pulchrum	Ericaceae	Woody	Evergreen	DAT5	4360	±	456	0.86	±	0.24
Rhododendron japonoheptamerum	Ericaceae	Woody	Evergreen	KOR4	2980	±	352	2.46	±	0.47
Rhododendron kaempferi	Ericaceae	Woody	Deciduous	KOR5	3200	±	126	1.72	±	0.34
Rhododendron sp.	Ericaceae	Woody	Evergreen	KAW5	10,300	±	1509	3.84	±	0.56
Rhododendron sp.	Ericaceae	Woody	Evergreen	SOM1	2530	±	90	3.21	±	0.15
Rhododendron sp.	Ericaceae	Woody	Evergreen	SOM1	1690	±	93	2.14	±	0.13
Rhus javanica	Anacardiaceae	Woody	Deciduous	KOR3	1520	±	79	0.38	±	0.05
Rhus javanica	Anacardiaceae	Woody	Deciduous	KAW4	396	±	15	0.12	±	0.00
Rhus trichocarpa	Anacardiaceae	Woody	Deciduous	KAW5	1560	±	41	0.58	±	0.02
Robinia pseudoacacia	Fabaceae	Woody	Deciduous	KOR3	379	±	50	0.09	±	0.02
Robinia pseudoacacia	Fabaceae	Woody	Deciduous	KOR2	N.D.	±	–	0.00	±	–
Rubus crataegifolius	Rosaceae	Woody	Deciduous	DAT1	5540	±	158	1.12	±	0.30
Rubus microphyllus	Rosaceae	Woody	Deciduous	NAR1	3320	±	94	1.12	±	0.20
Rubus microphyllus	Rosaceae	Woody	Deciduous	SOM1	558	±	44	0.71	±	0.06
Rubus microphyllus	Rosaceae	Woody	Deciduous	NAR1	2070	±	74	0.70	±	0.13
Rubus microphyllus	Rosaceae	Woody	Deciduous	SOM1	314	±	45	0.40	±	0.06
Rubus palmatus var. *coptophyllus*	Rosaceae	Woody	Deciduous	SOM1	5210	±	564	6.58	±	0.74
Rubus palmatus var. *coptophyllus*	Rosaceae	Woody	Deciduous	DAT5	15,100	±	266	2.96	±	0.78

(continued)

Sample name (Scientific name)	Family	Herbaceous/ woody	Lifeform	Sampling site	^{137}Cs concentration (Bq/kg DW)			CR value		
					Value	±	Error[a]	Value	±	Error[b]
Rubus palmatus var. *coptophyllus*	Rosaceae	Woody	Deciduous	DAT2	5180	±	53	1.19	±	0.20
Rubus palmatus var. *coptophyllus*	Rosaceae	Woody	Deciduous	KOR3	2570	±	117	0.64	±	0.09
Rubus palmatus var. *coptophyllus*	Rosaceae	Woody	Deciduous	KOR3	1550	±	91	0.39	±	0.06
Rubus palmatus var. *coptophyllus*	Rosaceae	Woody	Deciduous	DAT3	1310	±	14	0.24	±	0.04
Rubus palmatus var. *coptophyllus*	Rosaceae	Woody	Deciduous	DAT5	1060	±	133	0.21	±	0.06
Rubus palmatus var. *coptophyllus*	Rosaceae	Woody	Deciduous	DAT1	393	±	46	0.08	±	0.02
Rumex japonicus	Polygonaceae	Herbaceous	Perennial	DAT5	8890	±	991	1.74	±	0.50
Rumex japonicus	Polygonaceae	Herbaceous	Perennial	KOR1	N.D.	±	–	0.00	±	–
Rumex japonicus	Polygonaceae	Herbaceous	Perennial	KOR1	N.D.	±	–	0.00	±	–
Salix chaenomeloides	Salicaceae	Woody	Deciduous	DAT1	372	±	17	0.08	±	0.02
Salix integra	Salicaceae	Woody	Deciduous	KOR3	1930	±	98	0.48	±	0.07
Salix udensis	Salicaceae	Woody	Deciduous	KOR3	964	±	74	0.24	±	0.04
Sasamorpha borealis	Poaceae	Woody	Evergreen	SOM1	3240	±	80	4.11	±	0.16
Sasamorpha borealis	Poaceae	Woody	Evergreen	KAW1	9750	±	244	4.03	±	0.77
Sasamorpha borealis	Poaceae	Woody	Evergreen	KOR4	774	±	99	0.64	±	0.13
Setaria glauca	Poaceae	Herbaceous	Annual	KOR1	N.D.	±	–	0.00	±	–
Sicyos angulatus	Cucurbitaceae	Herbaceous	Annual	KOR2	393	±	45	0.07	±	0.02

Smilax china	Smilacaceae	Woody	Deciduous	KAW6	247	±	15	0.05	±	0.00
Solidago altissima	Asteraceae	Herbaceous	Perennial	FUT1	8120	±	128	0.11	±	0.02
Solidago altissima	Asteraceae	Herbaceous	Perennial	KOR3	288	±	49	0.07	±	0.02
Solidago altissima	Asteraceae	Herbaceous	Perennial	KOR1	238	±	54	0.04	±	0.01
Sonchus oleraceus	Asteraceae	Herbaceous	Perennial	HIR2	569	±	20	0.06	±	0.00
Sonchus oleraceus	Asteraceae	Herbaceous	Perennial	KAW4	131	±	12	0.04	±	0.00
Sonchus sp.	Asteraceae	Herbaceous	Perennial	KOR3	464	±	42	0.12	±	0.02
Sonchus sp.	Asteraceae	Herbaceous	Perennial	KAW4	N.D.	±	–	0.00	±	–
Stenactis annuus	Asteraceae	Herbaceous	Perennial	DAT5	1780	±	579	0.35	±	0.15
Stenactis annuus	Asteraceae	Herbaceous	Perennial	HIR2	787	±	24	0.09	±	0.00
Stenactis annuus	Asteraceae	Herbaceous	Perennial	KOR1	N.D.	±	–	0.00	±	–
Swida controversa	Cornaceae	Woody	Deciduous	KOR3	3470	±	123	0.86	±	0.12
Taraxacum sp.	Asteraceae	Herbaceous	Perennial	KAW4	2730	±	186	0.85	±	0.06
Trifolium pratense	Fabaceae	Herbaceous	Perennial	KOR3	2410	±	122	0.59	±	0.08
Trifolium pratense	Fabaceae	Herbaceous	Perennial	KOR1	294	±	15	0.05	±	0.01
Trifolium pratense	Fabaceae	Herbaceous	Perennial	KOR1	146	±	43	0.02	±	0.01
Trifolium repens	Fabaceae	Herbaceous	Perennial	SOM1	1180	±	124	1.48	±	0.16
Trifolium repens	Fabaceae	Herbaceous	Perennial	KOR1	816	±	137	0.13	±	0.03
Ulmus parvifolia	Ulmaceae	Woody	Deciduous	KAW4	2130	±	31	0.66	±	0.01

(continued)

Sample name (Scientific name)	Family	Herbaceous/ woody	Lifeform	Sampling site	^{137}Cs concentration (Bq/kg DW)			CR value		
					Value	±	Error[a]	Value	±	Error[b]
Viburnum furcatum	Caprifoliaceae	Woody	Deciduous	KAW5	3550	±	56	1.32	±	0.02
Viburnum odoratissimum var. *awabuk*	Caprifoliaceae	Woody	Evergreen	FUT1	10,900	±	187	0.14	±	0.03
Vicia angustifolia	Fabaceae	Herbaceous	Perennial	KOR2	774	±	97	0.13	±	0.04
Vicia angustifolia	Fabaceae	Herbaceous	Perennial	KOR1	148	±	13	0.02	±	0.00
Youngia denticulata	Asteraceae	Herbaceous	Annual	DAT2	N.D.	±	–	0.00	±	–

Appendix 2: Cesium-137 Concentration of Plant Leaves and Its Concentration Ratio (CR) in 2012

Sample name (scientific name)	Family	Herbaceous/ woody	Lifeform	Sampling site	^{137}Cs concentration (Bq/kg DW)			CR value		
					Value	±	Error[a]	Value	±	Error[b]
(Unknoun)	Asteraceae	Herbaceous	Perennial	KAW3	184	±	9	0.01	±	0.00
(Unknoun)	(Unknoun)	Herbaceous	Perennial	KOR1	70	±	3	0.01	±	0.00
(Unknoun)	Asteraceae	Herbaceous	Perennial	KOR1	N.D.	±	–	0.00	±	–
(Unknoun)	Poaceae	Herbaceous	Perennial	KAW2	N.D.	±	–	0.00	±	–
Acer crataegifolium	Aceraceae	Woody	Deciduous	KAW1	25,200	±	548	10.6	±	2.00
Acer crataegifolium	Aceraceae	Woody	Deciduous	KAW1	9140	±	127	3.86	±	0.73
Acer crataegifolium	Aceraceae	Woody	Deciduous	IWA2	2110	±	18	1.02	±	0.17
Acer rufinerve	Aceraceae	Woody	Deciduous	DAT3	879	±	13	0.16	±	0.03
Acer rufinerve	Aceraceae	Woody	Deciduous	DAT3	680	±	31	0.13	±	0.02
Acer shirasawanum	Aceraceae	Woody	Deciduous	KOR4	1110	±	38	0.93	±	0.14
Achillea millefolium	Asteraceae	Herbaceous	Perennial	DAT5	3132	±	24	0.63	±	0.17
Achillea millefolium	Asteraceae	Herbaceous	Perennial	DAT5	313	±	21	0.06	±	0.02
Achyranthes bidentata var. *tomentosa*	Amaranthaceae	Herbaceous	Perennial	KOR5	443	±	12	0.24	±	0.05

(continued)

Sample name (scientific name)	Family	Herbaceous/ woody	Lifeform	Sampling site	^{137}Cs concentration (Bq/kg DW)			CR value		
					Value	±	Error[a]	Value	±	Error[b]
Achyranthes bidentata var. *tomentosa*	Amaranthaceae	Herbaceous	Perennial	DAT2	390	±	11	0.09	±	0.02
Achyranthes bidentata var. *tomentosa*	Amaranthaceae	Herbaceous	Perennial	DAT5	204	±	8	0.04	±	0.01
Achyranthes bidentata var. *tomentosa*	Amaranthaceae	Herbaceous	Perennial	NAR1	81	±	4	0.03	±	0.01
Achyranthes bidentata var. *tomentosa*	Amaranthaceae	Herbaceous	Perennial	KAW2	174	±	7	0.03	±	0.00
Actinidia deliciosa	Actinidiaceae	Woody	Deciduous	FUT1	1930	±	50	0.03	±	0.01
Akebia quinata	Lardizabalaceae	Woody	Deciduous	KOR2	158	±	5	0.03	±	0.01
Alnus firma	Betulaceae	Woody	Deciduous	IWA2	2410	±	24	1.15	±	0.19
Alnus firma	Betulaceae	Woody	Deciduous	NAR1	169	±	14	0.06	±	0.01
Amaranthus lividus	Amaranthaceae	Herbaceous	Annual	SOM2	23	±	2	0.20	±	0.05
Amaranthus patulus	Amaranthaceae	Herbaceous	Annual	KAW2	51	±	3	0.01	±	0.00
Ambrosia trifida	Asteraceae	Herbaceous	Annual	KOR2	123	±	13	0.02	±	0.01
Ambrosia trifida	Asteraceae	Herbaceous	Annual	KOR2	100	±	6	0.02	±	0.01
Ambrosia trifida	Asteraceae	Herbaceous	Annual	KOR1	N.D.	±	–	0.00	±	–
Ampelopsis glandulosa var. *heterophylla*	Vitaceae	Woody	Deciduous	SOM1	35	±	4	0.05	±	0.00
Aralia elata	Araliaceae	Woody	Deciduous	DAT3	734	±	20	0.14	±	0.02

Aralia elata	Araliaceae	Woody	Deciduous	DAT3	210	±	13	0.04	±	0.01
Arisaema thunbergii spp. *urashima*	Araceae	Herbaceous	Perennial	KOR4	1340	±	63	1.13	±	0.18
Arisaema thunbergii spp. *urashima*	Araceae	Herbaceous	Perennial	KOR5	550	±	11	0.30	±	0.06
Artemisia indica var. *moximowiczii*	Asteraceae	Herbaceous	Perennial	NAR1	302	±	13	0.10	±	0.02
Artemisia indica var. *moximowiczii*	Asteraceae	Herbaceous	Perennial	KOR3	255	±	12	0.06	±	0.01
Artemisia indica var. *moximowiczii*	Asteraceae	Herbaceous	Perennial	KOR1	126	±	13	0.02	±	0.00
Artemisia indica var. *moximowiczii*	Asteraceae	Herbaceous	Perennial	KAW2	109	±	7	0.02	±	0.00
Bidens frondosa	Asteraceae	Herbaceous	Annual	HIR1	53	±	3	0.68	±	0.06
Bidens frondosa	Asteraceae	Herbaceous	Annual	KAW3	144	±	3	0.01	±	0.00
Boehmeria silvestrii	Urticaceae	Herbaceous	Perennial	KOR3	627	±	9	0.16	±	0.02
Boehmeria silvestrii	Urticaceae	Herbaceous	Perennial	DAT2	597	±	8	0.14	±	0.02
Boehmeria tricuspis	Urticaceae	Herbaceous	Perennial	SOM1	1530	±	35	1.97	±	0.07
Boehmeria tricuspis	Urticaceae	Herbaceous	Perennial	SOM1	369	±	8	0.48	±	0.02
Boehmeria tricuspis	Urticaceae	Herbaceous	Perennial	DAT2	541	±	24	0.13	±	0.02
Boehmeria tricuspis	Urticaceae	Herbaceous	Perennial	DAT1	491	±	10	0.10	±	0.03
Brassica juncea	Brassicaceae	Herbaceous	Annual	HIR1	N.D.	±	–	0.00	±	–
Brassica napus	Brassicaceae	Herbaceous	Annual	IWA1	68	±	10	0.18	±	0.08
Broussonetia kazinoki	Moraceae	Woody	Deciduous	SOM1	63	±	3	0.08	±	0.00

(continued)

Sample name (scientific name)	Family	Herbaceous/woody	Lifeform	Sampling site	^{137}Cs concentration (Bq/kg DW)			CR value		
					Value	±	Error[a]	Value	±	Error[b]
Callicarpa mollis	Verbenaceae	Woody	Deciduous	DAT1	351	±	8	0.07	±	0.02
Camellia japonica	Theaceae	Woody	Evergreen	KOR5	1029	±	8	0.56	±	0.11
Camellia sasanqua	Theaceae	Woody	Evergreen	KOR4	560	±	39	0.47	±	0.08
Camellia sasanqua	Theaceae	Woody	Evergreen	KOR5	292	±	9	0.16	±	0.03
Carpinus japonica	Betulaceae	Woody	Deciduous	SOM1	77	±	4	0.10	±	0.01
Carpinus tschonoskii	Betulaceae	Woody	Deciduous	SOM1	217	±	24	0.28	±	0.03
Castanea crenata	Fagaceae	Woody	Deciduous	DAT1	710	±	29	0.15	±	0.04
Castanea crenata	Fagaceae	Woody	Deciduous	DAT1	673	±	13	0.14	±	0.04
Cerasus jamasakura	Rosaceae	Woody	Deciduous	DAT2	188	±	11	0.04	±	0.01
Cerasus speciosa	Rosaceae	Woody	Deciduous	KOR3	239	±	4	0.06	±	0.01
Cerasus speciosa	Rosaceae	Woody	Deciduous	KOR3	106	±	14	0.03	±	0.00
Chengiopanax sciadophylloides	Araliaceae	Woody	Deciduous	KAW1	27,100	±	95	11.4	±	2.16
Chengiopanax sciadophylloides	Araliaceae	Woody	Deciduous	IWA2	6960	±	132	3.33	±	0.55
Chengiopanax sciadophylloides	Araliaceae	Woody	Deciduous	IWA2	4710	±	27	2.26	±	0.37
Chenopodium album	Chenopodiaceae	Herbaceous	Annual	SOM2	92	±	5	0.80	±	0.18
Chenopodium album	Chenopodiaceae	Herbaceous	Annual	HIR1	26	±	2	0.33	±	0.04

Chenopodium album	Chenopodiaceae	Herbaceous	Annual	SOM2	18	±	1	0.16	±	0.04
Chenopodium album	Chenopodiaceae	Herbaceous	Annual	IWA1	28	±	3	0.07	±	0.03
Chenopodium album	Chenopodiaceae	Herbaceous	Annual	HIR1	N.D.	±	–	0.00	±	–
Chenopodium ambrosioides	Chenopodiaceae	Herbaceous	Annual	HIR1	25	±	4	0.32	±	0.06
Chenopodium ambrosioides	Chenopodiaceae	Herbaceous	Annual	SOM2	N.D.	±	–	0.00	±	–
Chenopodium ficifolium	Chenopodiaceae	Herbaceous	Annual	HIR1	200	±	12	2.58	±	0.25
Chenopodium ficifolium	Chenopodiaceae	Herbaceous	Annual	SOM2	N.D.	±	–	0.00	±	–
Chenopodium ficifolium	Chenopodiaceae	Herbaceous	Annual	SOM2	N.D.	±	–	0.00	±	–
Citrus junos	Rutaceae	Woody	Evergreen	FUT1	7013	±	133	0.09	±	0.02
Clethra barbinervis	Clethraceae	Woody	Deciduous	IWA2	1920	±	22	0.93	±	0.15
Clethra barbinervis	Clethraceae	Woody	Deciduous	NAR1	430	±	34	0.15	±	0.03
Commelina sp.	Commelinaceae	Herbaceous	Annual	DAT5	295	±	7	0.06	±	0.02
Crypromeria japonica	Taxodiaceae	Woody	Evergreen	SOM1	444	±	22	0.58	±	0.03
Crypromeria japonica	Taxodiaceae	Woody	Evergreen	KOR3	1732	±	87	0.44	±	0.06
Crypromeria japonica	Taxodiaceae	Woody	Evergreen	SOM1	255	±	13	0.33	±	0.02
Crypromeria japonica	Taxodiaceae	Woody	Evergreen	KOR3	647	±	32	0.17	±	0.02
Diospyros kaki	Ebenaceae	Woody	Deciduous	FUT1	3203	±	53	0.04	±	0.01
Equisetum arvense	Equisetaceae	Herbaceous	Perennial	KOR3	88	±	6	0.02	±	0.00

(continued)

Sample name (scientific name)	Family	Herbaceous/ woody	Lifeform	Sampling site	^{137}Cs concentration (Bq/kg DW)			CR value		
					Value	±	Error[a]	Value	±	Error[b]
Equisetum arvense	Equisetaceae	Herbaceous	Perennial	KOR1	134	±	5	0.02	±	0.00
Equisetum arvense	Equisetaceae	Herbaceous	Perennial	KAW2	N.D.	±	–	0.00	±	–
Euonymus japonicus	Celastraceae	Woody	Evergreen	KOR5	359	±	10	0.20	±	0.04
Euptelea polyandra	Eupteleaceae	Woody	Deciduous	SOM1	51	±	3	0.07	±	0.00
Eurya japonica	Theaceae	Woody	Evergreen	KOR4	1270	±	51	1.07	±	0.17
Eurya japonica	Theaceae	Woody	Evergreen	IWA2	1466	±	58	0.70	±	0.12
Eurya japonica	Theaceae	Woody	Evergreen	KOR5	1210	±	49	0.66	±	0.13
Fraxinus sieboldiana	Oleaceae	Woody	Deciduous	KAW1	2990	±	70	1.25	±	0.24
Gamblea innovans	Araliaceae	Woody	Deciduous	IWA2	1950	±	26	0.94	±	0.15
Gamblea innovans	Araliaceae	Woody	Deciduous	IWA2	495	±	34	0.24	±	0.04
Hamamlis japonica	Hamamelidaceae	Woody	Deciduous	DAT3	1423	±	42	0.26	±	0.04
Hamamlis japonica	Hamamelidaceae	Woody	Deciduous	DAT3	465	±	7	0.09	±	0.01
Helianthus annuus	Asteraceae	Herbaceous	Annual	IWA1	222	±	6	0.57	±	0.25
Helianthus tuberosus	Asteraceae	Herbaceous	Perennial	DAT1	752	±	10	0.16	±	0.04
Helianthus tuberosus	Asteraceae	Herbaceous	Perennial	DAT1	551	±	25	0.11	±	0.03
Helianthus tuberosus	Asteraceae	Herbaceous	Perennial	KOR2	174	±	4	0.03	±	0.01

Houttuynia cordata	Saururaceae	Herbaceous	Perennial	KOR4	1520	±	51	1.28	±	0.20
Houttuynia cordata	Saururaceae	Herbaceous	Perennial	HIR1	90	±	14	1.16	±	0.21
Houttuynia cordata	Saururaceae	Herbaceous	Perennial	KOR5	1449	±	32	0.80	±	0.16
Houttuynia cordata	Saururaceae	Herbaceous	Perennial	NAR1	1135	±	54	0.39	±	0.07
Houttuynia cordata	Saururaceae	Herbaceous	Perennial	KOR3	413	±	28	0.10	±	0.02
Houttuynia cordata	Saururaceae	Herbaceous	Perennial	IWA1	38	±	10	0.10	±	0.05
Houttuynia cordata	Saururaceae	Herbaceous	Perennial	KOR3	229	±	8	0.06	±	0.01
Houttuynia cordata	Saururaceae	Herbaceous	Perennial	DAT4	169	±	16	0.04	±	0.01
Houttuynia cordata	Saururaceae	Herbaceous	Perennial	FUT1	585	±	38	0.01	±	0.00
Hydrangea macrophylla	Saxifragaceae	Woody	Deciduous	SOM1	695	±	22	0.89	±	0.04
Hydrangea macrophylla	Saxifragaceae	Woody	Deciduous	SOM1	275	±	6	0.36	±	0.01
Hydrangea sp.	Saxifragaceae	Woody	Deciduous	FUT1	2394	±	69	0.03	±	0.01
Ilex macropoda	Aquifoliaceae	Woody	Deciduous	KAW1	6730	±	42	2.84	±	0.54
Ilex macropoda	Aquifoliaceae	Woody	Deciduous	KAW1	4940	±	76	2.07	±	0.39
Ilex macropoda	Aquifoliaceae	Woody	Deciduous	SOM1	331	±	23	0.43	±	0.03
Lamium album var. *barbatum*	Lamiaceae	Herbaceous	Perennial	NAR1	2320	±	93	0.80	±	0.15
Ligustrum lucidum	Oleaceae	Woody	Evergreen	KOR5	948	±	6	0.52	±	0.10
Lilium auratum	Liliaceae	Herbaceous	Perennial	KOR4	1093	±	16	0.93	±	0.14

(continued)

Sample name (scientific name)	Family	Herbaceous/ woody	Lifeform	Sampling site	^{137}Cs concentration (Bq/kg DW)			CR value		
					Value	±	Error[a]	Value	±	Error[b]
Lilium auratum	Liliaceae	Herbaceous	Perennial	KOR5	747	±	33	0.41	±	0.08
Lindera umbellata	Lauraceae	Woody	Deciduous	DAT3	1864	±	33	0.35	±	0.06
Lindera umbellata	Lauraceae	Woody	Deciduous	DAT3	1189	±	17	0.22	±	0.04
Lithocarpus edulis	Fagaceae	Woody	Evergreen	IWA2	5810	±	133	2.78	±	0.46
Lysimachia clethroides	Primulaceae	Herbaceous	Perennial	KOR3	438	±	28	0.11	±	0.02
Lysimachia clethroides	Primulaceae	Herbaceous	Perennial	DAT1	212	±	7	0.04	±	0.01
Lysimachia clethroides	Primulaceae	Herbaceous	Perennial	DAT1	189	±	16	0.04	±	0.01
Magnolia kobus	Magnoliaceae	Woody	Deciduous	KOR3	1079	±	53	0.27	±	0.04
Mallotus japonicus	Euphorbiaceae	Woody	Deciduous	IWA2	918	±	25	0.44	±	0.07
Mentha arvensis var. *piperascens*	Lamiaceae	Herbaceous	Perennial	DAT5	1061	±	38	0.21	±	0.06
Mentha arvensis var. *piperascens*	Lamiaceae	Herbaceous	Perennial	DAT5	323	±	7	0.07	±	0.02
Miscanthus sinensis	Poaceae	Herbaceous	Perennial	DAT1	444	±	7	0.09	±	0.02
Miscanthus sinensis	Poaceae	Herbaceous	Perennial	NAR1	110	±	10	0.04	±	0.01
Miscanthus sinensis	Poaceae	Herbaceous	Perennial	DAT1	163	±	11	0.03	±	0.01
Miscanthus sinensis	Poaceae	Herbaceous	Perennial	KAW2	192	±	30	0.03	±	0.01

Miscanthus sinensis	Poaceae	Herbaceous	Perennial	KAW2	158	±	47	0.02	±	0.01
Miscanthus sinensis	Poaceae	Herbaceous	Perennial	KAW2	105	±	27	0.02	±	0.00
Miscanthus sinensis	Poaceae	Herbaceous	Perennial	NAR1	44	±	2	0.02	±	0.00
Miscanthus sinensis	Poaceae	Herbaceous	Perennial	KOR2	24	±	1	0.00	±	0.00
Monochoria vaginalis var. *plantaginea*	Pontederiaceae	Herbaceous	Annual	DAT4	N.D.	±	–	0.00	±	–
Murdannia keisak	Commelinaceae	Herbaceous	Annual	DAT4	182	±	11	0.05	±	0.01
Nicotiana tabacum	Solanaceae	Herbaceous	Perennial	KAW3	395	±	21	0.03	±	0.01
Oenothera biennis	Onagraceae	Herbaceous	Perennial	NAR1	14,700	±	50	5.04	±	0.90
Oenothera biennis	Onagraceae	Herbaceous	Perennial	KAW2	71	±	2	0.01	±	0.00
Oryza sativa	Poaceae	Herbaceous	Annual	DAT4	N.D.	±	–	0.00	±	–
Osmanthus × fortunei	Oleaceae	Woody	Evergreen	FUT1	6056	±	139	0.08	±	0.02
Osmunda japonica	Osmundaceae	Herbaceous	Perennial	KOR4	866	±	11	0.73	±	0.11
Perilla frutescens	Lamiaceae	Herbaceous	Annual	IWA1	1	±	0	0.00	±	0.00
Persicaria lapathifolia	Polygonaceae	Herbaceous	Annual	HIR1	32	±	2	0.42	±	0.05
Persicaria lapathifolia	Polygonaceae	Herbaceous	Annual	KOR1	116	±	3	0.02	±	0.00
Persicaria lapathifolia	Polygonaceae	Herbaceous	Annual	KAW3	119	±	4	0.01	±	0.00
Petasites japonicus	Asteraceae	Herbaceous	Perennial	DAT2	735	±	11	0.17	±	0.03
Petasites japonicus	Asteraceae	Herbaceous	Perennial	DAT2	463	±	10	0.11	±	0.02

(continued)

Sample name (scientific name)	Family	Herbaceous/ woody	Lifeform	Sampling site	^{137}Cs concentration (Bq/kg DW)			CR value		
					Value	±	Error[a]	Value	±	Error[b]
Petasites japonicus	Asteraceae	Herbaceous	Perennial	KOR3	386	±	15	0.10	±	0.01
Petasites japonicus	Asteraceae	Herbaceous	Perennial	KOR3	356	±	26	0.09	±	0.01
Petasites japonicus	Asteraceae	Herbaceous	Perennial	KOR3	151	±	5	0.04	±	0.01
Photinia glabra	Rosaceae	Woody	Evergreen	NAR1	1034	±	46	0.35	±	0.07
Photinia glabra	Rosaceae	Woody	Evergreen	NAR1	940	±	12	0.32	±	0.06
Pieris japonica	Ericaceae	Woody	Evergreen	IWA2	957	±	69	0.46	±	0.08
Pieris japonica	Ericaceae	Woody	Evergreen	KOR5	378	±	16	0.21	±	0.04
Pinus thunbergii	Pinaceae	Woody	Evergreen	NAR1	7320	±	196	2.51	±	0.45
Plantago asiatica	Plantaginaceae	Herbaceous	Perennial	KAW3	164	±	18	0.01	±	0.00
Plantago asiatica	Plantaginaceae	Herbaceous	Perennial	KOR1	21	±	1	0.00	±	0.00
Polygonum sp.	Polygonaceae	Herbaceous	Annual	KOR1	380	±	14	0.06	±	0.01
Polygonum thunbergii	Polygonaceae	Herbaceous	Annual	KAW3	557	±	10	0.04	±	0.01
Polygonum thunbergii	Polygonaceae	Herbaceous	Annual	KAW3	49	±	2	0.00	±	0.00
Populus tremula var. *sieboldii*	Salicaceae	Woody	Deciduous	KOR3	1967	±	66	0.50	±	0.07
Pteridium aquilinum	Dennstaedtiaceae	Herbaceous	Perennial	KOR3	8990	±	72	2.27	±	0.30

Pteridium aquilinum	Dennstaedtiaceae	Herbaceous	Perennial	DAT2	60	±	6	0.01	±	0.00
Pteridium aquilinum	Dennstaedtiaceae	Herbaceous	Perennial	DAT2	46	±	2	0.01	±	0.00
Pteridium aquilinum	Dennstaedtiaceae	Herbaceous	Perennial	KOR2	30	±	4	0.01	±	0.00
Pueraria lobata	Fabaceae	Herbaceous	Perennial	IWA1	30	±	2	0.08	±	0.03
Pueraria lobata	Fabaceae	Herbaceous	Perennial	KOR3	212	±	5	0.05	±	0.01
Pueraria lobata	Fabaceae	Herbaceous	Perennial	KOR2	67	±	13	0.01	±	0.00
Quercus aliena	Fagaceae	Woody	Deciduous	KAW1	8590	±	143	3.60	±	0.68
Quercus crispula	Fagaceae	Woody	Deciduous	DAT3	2350	±	51	0.44	±	0.07
Quercus crispula	Fagaceae	Woody	Deciduous	DAT3	1146	±	11	0.21	±	0.03
Quercus serrata	Fagaceae	Woody	Deciduous	IWA2	1756	±	51	0.84	±	0.14
Quercus serrata	Fagaceae	Woody	Deciduous	DAT3	1628	±	57	0.30	±	0.05
Ranunculus cantoniensis	Ranunculaceae	Herbaceous	Perennial	KOR1	38	±	2	0.01	±	0.00
Raphanus sativus var. *longipinnatus*	Brassicaceae	Herbaceous	Perennial	IWA1	90	±	13	0.23	±	0.11
Reynoutria japonica	Polygonaceae	Herbaceous	Perennial	IWA1	30	±	2	0.08	±	0.03
Reynoutria japonica	Polygonaceae	Herbaceous	Perennial	KOR2	110	±	6	0.02	±	0.01
Rhododendron japonoheptamerum	Ericaceae	Woody	Evergreen	KOR4	1029	±	51	0.87	±	0.14
Rhododendron sp.	Ericaceae	Woody	Evergreen	SOM1	765	±	11	0.98	±	0.03
Rhododendron sp.	Ericaceae	Woody	Evergreen	NAR1	1171	±	15	0.40	±	0.07

(continued)

Sample name (scientific name)	Family	Herbaceous/ woody	Lifeform	Sampling site	^{137}Cs concentration (Bq/kg DW)			CR value		
					Value	±	Error[a]	Value	±	Error[b]
Rhus javanica	Anacardiaceae	Woody	Deciduous	KOR3	452	±	24	0.11	±	0.02
Robinia pseudoacacia	Fabaceae	Woody	Deciduous	KOR2	31	±	3	0.01	±	0.00
Rubus crataegifolius	Rosaceae	Woody	Deciduous	DAT1	553	±	13	0.11	±	0.03
Rubus crataegifolius	Rosaceae	Woody	Deciduous	DAT1	169	±	15	0.03	±	0.01
Rubus microphyllus	Rosaceae	Woody	Deciduous	NAR1	137	±	5	0.05	±	0.01
Rubus microphyllus	Rosaceae	Woody	Deciduous	NAR1	126	±	15	0.04	±	0.01
Rubus palmatus var. *coptophyllus*	Rosaceae	Woody	Deciduous	IWA2	280	±	25	0.13	±	0.03
Rubus palmatus var. *coptophyllus*	Rosaceae	Woody	Deciduous	KOR3	470	±	23	0.12	±	0.02
Rubus palmatus var. *coptophyllus*	Rosaceae	Woody	Deciduous	DAT3	196	±	8	0.04	±	0.01
Rubus palmatus var. *coptophyllus*	Rosaceae	Woody	Deciduous	DAT2	156	±	11	0.04	±	0.01
Rubus palmatus var. *coptophyllus*	Rosaceae	Woody	Deciduous	KOR3	50	±	3	0.01	±	0.00
Rubus palmatus var. *coptophyllus*	Rosaceae	Woody	Deciduous	SOM1	N.D.	±	–	0.00	±	–
Rumex japonicus	Polygonaceae	Herbaceous	Perennial	KAW3	223	±	5	0.01	±	0.00
Rumex japonicus	Polygonaceae	Herbaceous	Perennial	KOR1	80	±	4	0.01	±	0.00
Rumex japonicus	Polygonaceae	Herbaceous	Perennial	KAW2	89	±	3	0.01	±	0.00
Rumex japonicus	Polygonaceae	Herbaceous	Perennial	KAW2	57	±	5	0.01	±	0.00

Rumex japonicus	Polygonaceae	Herbaceous	Perennial	KAW2	57	±	4	0.01	±	0.00
Rumex japonicus	Polygonaceae	Herbaceous	Perennial	KAW2	54	±	5	0.01	±	0.00
Rumex japonicus	Polygonaceae	Herbaceous	Perennial	KAW3	88	±	10	0.01	±	0.00
Rumex japonicus	Polygonaceae	Herbaceous	Perennial	IWA1	N.D.	±	–	0.00	±	–
Sicyos angulatus	Cucurbitaceae	Herbaceous	Annual	KOR2	283	±	5	0.05	±	0.01
Smilax china	Smilacaceae	Woody	Deciduous	KAW1	445	±	16	0.19	±	0.04
Solidago altissima	Asteraceae	Herbaceous	Perennial	NAR1	6360	±	127	2.19	±	0.39
Solidago altissima	Asteraceae	Herbaceous	Perennial	KOR3	320	±	7	0.08	±	0.01
Solidago altissima	Asteraceae	Herbaceous	Perennial	KAW2	372	±	37	0.05	±	0.01
Solidago altissima	Asteraceae	Herbaceous	Perennial	KAW2	338	±	46	0.05	±	0.01
Solidago altissima	Asteraceae	Herbaceous	Perennial	KAW2	264	±	32	0.04	±	0.01
Solidago altissima	Asteraceae	Herbaceous	Perennial	FUT1	1990	±	26	0.03	±	0.01
Solidago altissima	Asteraceae	Herbaceous	Perennial	FUT1	1447	±	40	0.02	±	0.00
Solidago altissima	Asteraceae	Herbaceous	Perennial	KOR1	56	±	5	0.01	±	0.00
Solidago altissima	Asteraceae	Herbaceous	Perennial	KAW2	34	±	4	0.00	±	0.00
Sonchus oleraceus	Asteraceae	Herbaceous	Perennial	DAT1	323	±	8	0.07	±	0.02
Stenactis annuus	Asteraceae	Herbaceous	Perennial	IWA1	N.D.	±	–	0.00	±	–
Swida controversa	Cornaceae	Woody	Deciduous	KOR3	467	±	22	0.12	±	0.02

(continued)

Sample name (scientific name)	Family	Herbaceous/ woody	Lifeform	Sampling site	^{137}Cs concentration (Bq/kg DW)			CR value		
					Value	±	Error[a]	Value	±	Error[b]
Taraxacum sp.	Asteraceae	Herbaceous	Perennial	FUT1	2158	±	46	0.03	±	0.01
Taraxacum sp.	Asteraceae	Herbaceous	Perennial	FUT1	820	±	22	0.01	±	0.00
Trifolium pratense	Fabaceae	Herbaceous	Perennial	KAW2	115	±	10	0.02	±	0.00
Trifolium repens	Fabaceae	Herbaceous	Perennial	KOR1	131	±	20	0.02	±	0.00
Trifolium repens	Fabaceae	Herbaceous	Perennial	KAW2	109	±	5	0.02	±	0.00
Ulmus davidiana var. *japonica*	Ulmaceae	Woody	Deciduous	KAW1	4350	±	90	1.83	±	0.35
Viburnum furcatum	Caprifoliaceae	Woody	Deciduous	KAW1	5310	±	29	2.24	±	0.42
Viburnum furcatum	Caprifoliaceae	Woody	Deciduous	KAW1	2550	±	70	1.07	±	0.20
Wisteria floribunda	Fabaceae	Woody	Deciduous	SOM1	1640	±	59	2.11	±	0.10

References

Broadley MR, Willey NJ (1997) Differences in root uptake of radiocaesium by 30 plant taxa. Environ Pollut 97:11–15

Broadley M, Willey NJ, Mead A (1999) A method to assess taxonomic variation in shoot caesium concentration among flowering plants. Environ Pollut 106:341–349

Chino M, Nakayama H, Nagai H, Terada H, Katata G, Yamazawa H (2011) Preliminary estimation of release amounts of ^{131}I and ^{137}Cs accidentally discharged from the Fukushima Daiichi nuclear power plant into the atmosphere. J Nucl Sci Technol 48:1129–1134

Deguchi S, Matsuda Y, Takenaka C, Sugiura Y, Ozawa H, Ogata Y (2017) Proposal of a new method of colonization rate or arbuscular mycorrhizal fungi in the root of *Chengiopanax sciadophylloides*. Mycobio 45(1):15–19

Dushenkov S, Mikheev A, Prokhnevsky A, Ruchko M, Sorochinsky B (1999) Phytoremediation of radiocesium-contaminated soil in the vicinity of Chernobyl, Ukraine. Environ Sci Technol 33:469–475

Ehlken S, Kirchner G (2002) Environmental processes affecting plant root uptake of radioactive trace elements and variability of transfer factor data: a review. J Environ Radioact 58:97–112

Forest Agency (2014) Forest products and radioactive materials – toward the revival of forest and forestry of Fukushima – [published in Japanese]. For For Prod Res Inst

Fuhrmann M, Lasat M, Ebbs S, Cornish J, Kochian L (2003) Uptake and release of cesium-137 by five plant species as influenced by soil amendments in field experiments. J Environ Qual 32:2272–2279

Hasegawa H, Tsukada H, Kawabata H, Chikuchi Y, Takaku Y, Hisamatsu S (2009) Effect of the counter anion of cesium on foliar uptake and translocation. J Environ Radioact 100:54–57

Hirose K (2016) Fukushima Daiichi nuclear plant accident: atmospheric and oceanic impacts over the five years. J Environ Radioact 157:113–130

Iijima K (2015) Status of the researches on the behavior in the environment of radioactive cesium transported from forests to river systems. Chikyukagaku (Geochemistry) 49:203–215

JAEA (2015) Long-term assessment of transport of radioactive contaminant in the environment of Fukushima [published in Japanese] [WWW Document]. Japan At Energy Agency. http://fukushima.jaea.go.jp/initiatives/cat01/pdf/project.pdf

Kamei-Ishikawa N, Uchida S, Tagami K (2008) Distribution coefficients for ^{85}Sr and ^{137}Cs in Japanese agricultural soils and their correlations with soil properties. J Radioanal Nucl Chem 277:433–439

Kanasashi T, Sugiura Y, Takenaka C, Hijii N, Umemura M (2015) Radiocesium distribution in sugi (*Cryptomeria japonica*) in Eastern Japan: translocation from needles to pollen. J Environ Radioact 139:398–406

Kitamura A, Yamaguchi M, Kurikami H, Yui M, Onishi Y (2014) Predicting sediment and cesium-137 discharge from catchments in eastern Fukushima. Anthropocene 5:22–31

Kobayashi T, Nagai H, Chino M, Kawamura H (2013) Source term estimation of atmospheric release due to the Fukushima Dai-ichi nuclear power plant accident by atmospheric and oceanic dispersion simulations. J Nucl Sci Technol 50:255–264

Lasat MM, Fuhrmann M, Ebbs SD, Cornish JE, Kochian LV (1998) Phytoremediation of a radiocesium-contaminated soil: evaluation of cesium-137 bioaccumulation in the shoots of three plant species. J Environ Qual 27:165–169

Memon AR, Itô S, Yatazawa M (1979) Absorption and accumulation of iron, manganese and copper in plants in the temperate forest of Central Japan. Soil Sci Plant Nutr 25:611–620

MHLW (2015) Information on the great east Japan earthquake [WWW Document]. Minist. Heal. Welf. Labor. https://www.mhlw.go.jp/shinsai_jouhou/shokuhin.html

Mimura T, Mimura M, Kobayashi D, Komiyama C, Sekimoto H, Miyamoto M, Kitamura A (2014) Radioactive pollution and accumulation of radionuclides in wild plants in Fukushima. J Plant Res 127:5–10

Mizuno T, Asahina R, Hosono A, Tanaka A, Senoo K, Obata H (2008) Age-dependent manganese hyperaccumulation in *Chengiopanax sciadophylloides* (Araliaceae). J Plant Nutr 31:1811–1819
NRA (2014) Monitoring information of environmental radioactivity level [published in Japanese]. Nucl Regul Soc
Sakamoto F, Ohnuki T, Kozai N, Igarashi S, Yamasaki S, Yoshida Z, Tanaka S (2012) Local area distribution of fallout radionuclides from Fukushima Daiichi nuclear power plant determined by autoradiography analysis. Trans At Energy Soc Japan 11:1–7
Sawidis T, Drossos E, Heinrich G, Papastefanou C (1990) Cesium-137 accumulation in higher plants before and after Chernobyl. Environ Int 16:163–169
Sugiura Y, Kanasashi T, Ogata Y, Ozawa H, Takenaka C (2016) Radiocesium accumulation properties of *Chengiopanax sciadophylloides*. J Environ Radioact 151:250–257
Tagami K, Uchida S, Ishii N, Kagiya S (2012) Translocation of radiocesium from stems and leaves of plants and the effect on radiocesium concentrations in newly emerged plant tissues. J Environ Radioact 111:65–69
Takeda A, Tsukada H, Nakao A, Takaku Y, Hisamatsu S (2013) Time-dependent changes of phytoavailability of Cs added to allophanic andosols in laboratory cultivations and extraction tests. J Environ Radioact 122:29–36
Tsukada H, Hasegawa H (2002) Soil-to-plant transfer of ^{137}Cs and other essential and trace elements in cabbage plants. J Radioanal Nucl Chem 252:219–224
Tsukada H, Nakamura Y (1999) Transfer of ^{137}Cs and stable Cs from soil to potato in agricultural fields. Sci Total Environ 228:111–120
Vinichuk M, Mårtensson A, Ericsson T, Rosén K (2013a) Effect of arbuscular mycorrhizal (AM) fungi on ^{137}Cs uptake by plants grown on different soils. J Environ Radioact 115:151–156
Vinichuk M, Mårtensson A, Rosén K (2013b) Inoculation with arbuscular mycorrhizae does not improve ^{137}Cs uptake in crops grown in the Chernobyl region. J Environ Radioact 126:14–19
Yamaguchi N, Takada Y, Hayashi K, Ishikawa S, Kuramata M, Eguchi S, Yoshikawa S, Sakaguchi A, Asada K, Wagai R, Makino T, Akahane I, Hiradate S (2012) Behavior of radiocaesium in soil-plant and its controlling factor [published in Japanese]. Bull Natl Inst Agro-Environmental Sci 31:75–129
Yamaji K, Nagata S, Haruma T, Ohnuki T, Kozaki T, Watanabe N, Nanba K (2016) Root endophytic bacteria of a ^{137}Cs and Mn accumulator plant, Eleutherococcus sciadophylloides, increase ^{137}Cs and Mn desorption in the soil. J Environ Radioact 153:112–119
Yamashita J, Enomoto T, Yamada M, Ono T, Hanafusa T, Nagamatsu T, Sonoda S, Yamamoto Y (2014) Estimation of soil-to-plant transfer factors of radiocesium in 99 wild plant species grown in arable lands 1 year after the Fukushima 1 nuclear power plant accident. J Plant Res 127:11–22
Zhu YG, Smolders E (2000) Plant uptake of radiocaesium: a review of mechanisms, regulation and application. J Exp Bot 51:1635–1645

Chapter 4
Surface Absorption of ^{137}Cs Through Tree Bark

Wei Wang

Abstract Bark–wood translocation has been suggested as an important route for the contamination of trees via radiocesium in the initial stage of deposition. We investigated cesium absorption through the bark of Japanese cedar (*Cryptomeria japonica*), konara oak (*Quercus serrata*), and red pine (*Pinus densiflora*). Stable Cs (^{133}Cs) was applied to the bark at 1.2-m height of the three tree species. The ^{133}Cs concentrations were determined in the bark, sapwood, and heartwood of stem disks from several heights, as well as in current-year foliage/needles from the canopy. The results suggested that Cs can enter trees through the bark for all the three tree species even though in the dormant period of tree, and the absorption is probably independent of cambial activities. The distribution pattern of Cs within trees varies among different species.

Keywords Bark–wood translocation · Contamination of trees · Radiocesium · Stable cesium (^{133}Cs)

4.1 Introduction

The Fukushima Daiichi Nuclear Power Plant (FDNPP) accident in March 2011 resulted in a considerable release of radiocesium, which contaminated a wide area of the Fukushima Prefecture (Hashimoto et al. 2012), especially forests, which cover approximately 70% of the entire prefectural area. Because radiocesium can be trapped and recycled in forest ecosystems, and decontamination of radiocesium from contaminated forests is difficult, the radioactive contamination of forests and forest products and the long-term effects of radionuclides on the forest biota and humans are of great concern (Aliyu et al. 2015; Hasegawa et al. 2013).

W. Wang (✉)
Shanghai Institute of Applied Physics, Chinese Academy of Sciences, Shanghai, China
e-mail: wangwei1986@sinap.ac.cn

C. Takenaka et al. (eds.), *Radiocesium Dynamics in a Japanese Forest Ecosystem*,
https://doi.org/10.1007/978-981-13-8606-0_4

Studies on the migration and distribution of ^{137}Cs within forests in the early stage of contamination are essential to estimate its long-term fate (Kato et al. 2012; Endo et al. 2015). One of the important concerns is the route by which ^{137}Cs is absorbed by trees. Three potential pathways of ^{137}Cs absorption by trees need to be considered, i.e., surface absorption through leaves, bark, and root uptake from the soil; however, which pathway is the primary route is not well understood. Several factors, such as tree species, soil type, chemical speciation of the ^{137}Cs species, bioavailability, and the distribution of ^{137}Cs in the soil profile, complicate the absorption process. Although previous studies have stated the possible uptake pathways, including foliar absorption (Calmon et al. 2009; Tagami et al. 2012; Nishikiori et al. 2015) or absorption through the bark (Mahara et al. 2014; Sato et al. 2015; Sugiura et al. 2016), only little experimental demonstration has been made (Ohta et al. 2016). For future predictions of radioactive contamination in forests, it is important to elucidate how ^{137}Cs is taken up by trees at the initial stage of deposition and what is the dynamic behavior of ^{137}Cs internally (Nishikiori et al. 2015).

Our primary aim was to determine whether Cs could move into a tree stem through the bark and to assess how Cs moves within different tree species. As the chemical behavior of ^{137}Cs is expected to be almost identical to that of ^{133}Cs, analysis of ^{133}Cs is useful for understanding the behavior of ^{137}Cs in a forest ecosystem (Yoshida et al. 2004). In the present study, Japanese cedar (*Cryptomeria japonica*), konara oak (*Quercus serrata*), and red pine (*Pinus densiflora*) were selected, and ^{133}Cs was applied to the bark of the three trees species under field conditions to follow its movement in the tree.

4.2 Stable Cesium Concentration in Different Tree Parts

To check whether the soils under the tree stems were contaminated with ^{133}Cs, we measured the ion-exchangeable ^{133}Cs concentration in the surface soils under both the ^{133}Cs-treated trees and the control trees (Table 4.1). No significant differences were observed between the average ^{133}Cs concentration in the soils under the treated trees and those under the control trees, indicating that the influence of root uptake caused by the leakage of the $^{133}CsCl$ solution was negligible.

Table 4.1 Ion-exchangeable ^{133}Cs concentrations in surface soils under ^{133}Cs-treated trees and control trees ($n = 3$)

Trees	Treated (μg g^{-1})	Control (μg g^{-1})
10-year-old *C. japonica*	0.021 ± 0.0090	0.033 ± 0.019
26-year-old *C. japonica*	0.12 ± 0.041	0.13 ± 0.022
Q. serrata	0.087 ± 0.021	0.065 ± 0.039
P. densiflora	0.11 ± 0.039	0.010 ± 0.0013

Table 4.2 ^{133}Cs (μg g^{-1}) and K concentrations in needles, bark, and sapwood of disks sampled at various heights from 10-year-old *C. japonica*

Component	Position	KT		KC	
		Cs (μg g^{-1})	K (mg g^{-1})	Cs (μg g^{-1})	K (mg g^{-1})
Needles	Tree top	19 ± 7.9	22 ± 1.5	0.20 ± 0.082	26 ± 5.2
	2.2 m	9.3 ± 5.9	5.2 ± 0.21	–	–
Bark	1.2 m	58 ± 30	4.5 ± 0.25	0.020 ± 0.0058	2.7 ± 0.60
	0.2 m	54 ± 30	5.1 ± 0.21	–	–
	2.2 m	2.6 ± 1.9	1.3 ± 0.088	–	–
Sapwood	1.2 m	8.0 ± 3.6	1.1 ± 0.15	0.013 ± 0.0033	1.2 ± 0.12
	0.2 m	1.7 ± 0.59	0.99 ± 0.12	–	–

Concentrations are average ± SE of three trees. Disk samples were collected from 0.2 m, 1.2 m to 2.2 m, while needles were sampled from tree top. For control trees, concentration values are only shown for 1.2-m disks

KT mean of mean ^{133}Cs-treated trees in Koriyama, KC mean of control trees in Koriyama

4.2.1 ^{133}Cs and K in Bark, Wood, and Needles of 10-Year-Old C. japonica

The concentrations of ^{133}Cs and K in the sapwood at three vertical positions as well as in the needles of 10-year-old cedars are shown in Table 4.2. At 1.2 m, the ^{133}Cs concentrations in the sapwood of the ^{133}Cs-treated trees were significantly higher than in the control trees ($p < 0.05$). Also, the average ^{133}Cs concentrations in needles from ^{133}Cs-treated trees was much higher than in the control trees though there were no significant differences between them. No significant differences were observed for average K concentrations in sapwood at 1.2 m for treated trees (1.1 mg g^{-1}) and control trees (1.2 mg g^{-1}); also no significant differences were observed for average K concentrations in needles of treated trees (22 mg g^{-1}) and control trees (26 mg g^{-1}). Figure 4.1 shows the vertical distribution of the average ^{133}Cs concentrations in the bark and sapwood of 10-year-old ^{133}Cs-treated trees. The average ^{133}Cs concentrations in the bark and the sapwood at various vertical positions were in the order of 1.2 > 0.2 > 2.2 m and 1.2 > 2.2 > 0.2 m, respectively.

4.2.2 ^{133}Cs and K in Bark, Wood, and Needles of 26-Year-Old C. japonica

The concentrations of ^{133}Cs and K in the bark, sapwood, and heartwood at different heights and in the needles of 26-year-old *C. japonica* are presented in Table 4.3. At 1.2 m, the ^{133}Cs concentrations in sapwood and heartwood of ^{133}Cs-treated trees were considerably higher than those of control trees, although there were no significant differences among them. In addition, the ^{133}Cs concentrations in the heartwood tended to be higher than in the sapwood for both ^{133}Cs-treated and

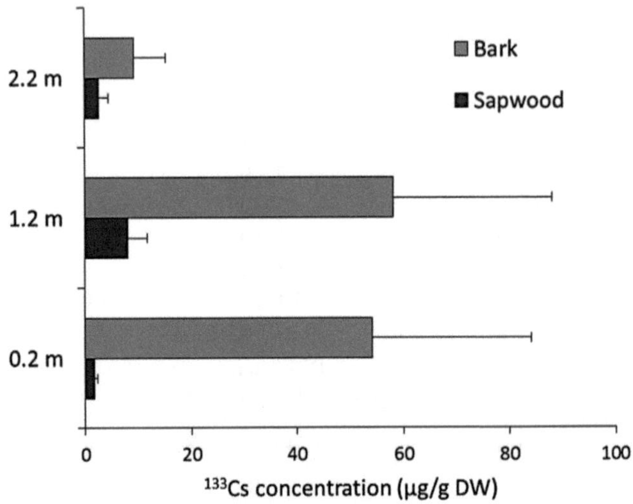

Fig. 4.1 The concentration profiles of ^{133}Cs in the bark and sapwood of the ^{133}Cs-treated 10-year-old *C. japonica*. Error bars indicate the SE ($n = 3$)

Table 4.3 ^{133}Cs (μg g^{-1}) and K concentrations in needles, bark, sapwood, and heartwood of disks sampled at various heights from 26-year-old *C. japonica*

Component	Position	UT		UC	
		Cs (μg g^{-1})	K (mg g^{-1})	Cs (μg g^{-1})	K (mg g^{-1})
Needles	Tree top	0.59 ± 0.30	6.2 ± 0.82	0.099 ± 0.036	4.7 ± 0.21
	6.2 m	0.32 ± 0.14	1.1 ± 0.33	–	–
Bark	1.2 m	18 ± 3.2	1.1 ± 0.27	0.037 ± 0.0033	0.89 ± 0.17
	0.2 m	1.6 ± 1.1	1.4 ± 0.23	–	–
	6.2 m	0.33 ± 0.27	0.68 ± 0.063	–	–
Sapwood	1.2 m	2.1 ± 0.55	0.59 ± 0.090	0.10 ± 0.078	0.56 ± 0.087
	0.2 m	0.38 ± 0.20	0.89 ± 0.16	–	–
	6.2 m	0.25 ± 0.16	3.4 ± 0.24	–	–
Heartwood	1.2 m	3.6 ± 0.87	1.6 ± 0.32	0.33 ± 0.25	1.4 ± 0.26
	0.2 m	0.39 ± 0.10	2.4 ± 0.22	–	–

Concentrations are average ± SE of three trees. Disk samples were collected from 0.2 m, 1.2 m to 6.2 m, while needles were sampled from tree top. For control trees, concentration values are only shown for 1.2-m disks. UT mean of mean ^{133}Cs-treated trees in Utsunomiya, UC mean of control trees in Utsunomiya

control trees. The average concentrations of ^{133}Cs in needles of the treated trees (0.59 μg g^{-1}) were significantly higher than those of the control trees (0.099 μg g^{-1}) ($p < 0.05$). No significant difference was observed for average K concentrations in needles, sapwood, and heartwood for the 1.2-m disk for treated trees and control

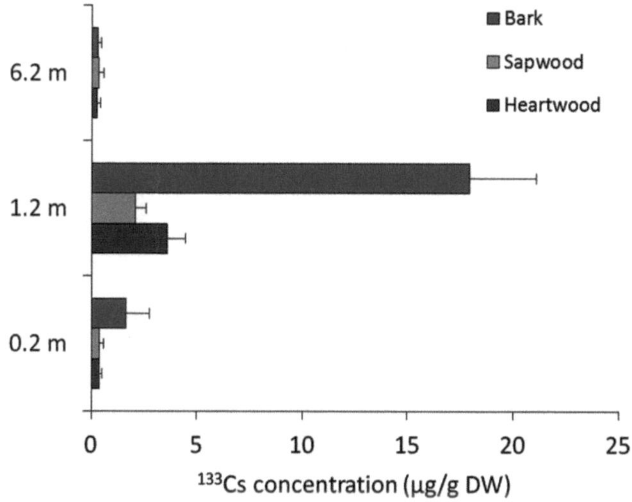

Fig. 4.2 The concentration profiles of ^{133}Cs in the bark, sapwood, and heartwood of the ^{133}Cs-treated 26-year-old *C. japonica*. Error bars indicate the SE ($n = 3$)

trees. The distribution of ^{133}Cs concentrations in ^{133}Cs-treated trees at Utsunomiya is shown in Fig. 4.2. The average ^{133}Cs concentrations at 1.2 m were higher than at 0.2 and 6.2 m.

4.2.3 ^{133}Cs in Bark, Wood, and Leaves of Q. serrata

Figure 4.3 shows the radial distribution of the ^{133}Cs concentration in the bark and wood at different vertical heights. Disks at 1/2 the tree height and 3/4 the tree height were combined together as the "upper parts" and disks at 0.2 m were considered to be "lower parts." At 1.2 m, the average ^{133}Cs concentration in the sapwood of the ^{133}Cs-treated trees was much higher than those of the control trees, even though there were no significant differences between them. The ^{133}Cs concentration decreased toward the heartwood with increasing distance from the outer bark. No significant difference was observed in the ^{133}Cs concentration in the heartwood between the ^{133}Cs-treated trees and the control trees. In both the lower and upper parts, no significant difference was observed for the ^{133}Cs concentration in both the sapwood and the heartwood between the ^{133}Cs-treated trees and the control trees. The ^{133}Cs concentrations in the leaves and branches of the treated trees (0.038 ± 0.016 µg g^{-1} and 0.040 ± 0.016 µg g^{-1}, respectively) were lower than those of the control trees (0.090 ± 0.018 µg g^{-1} and 0.074 ± 0.028 µg g^{-1}, respectively).

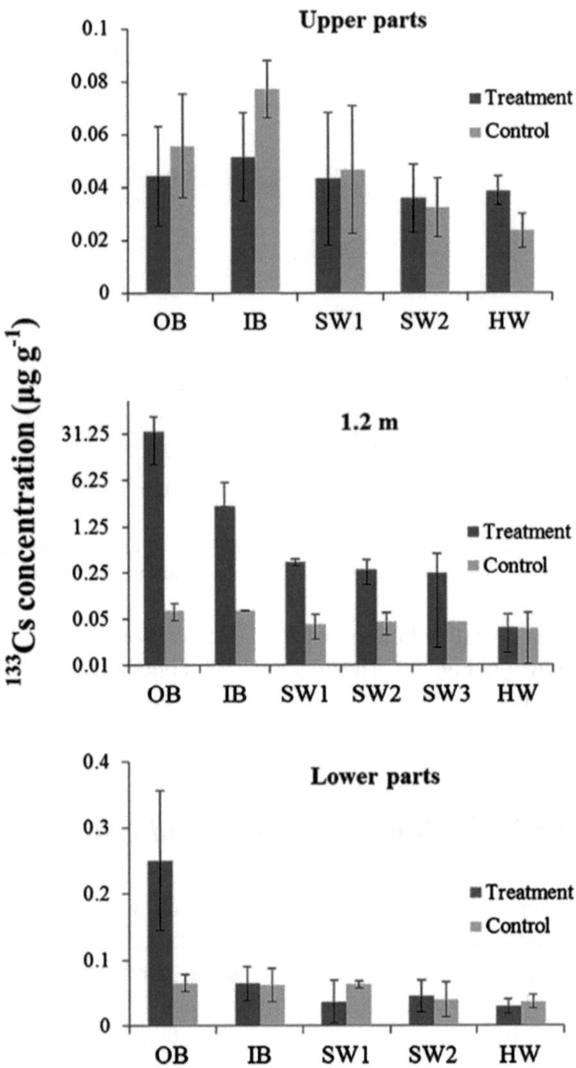

Fig. 4.3 Radial distribution of the ^{133}Cs concentration (μg g^{-1}) in the outer bark (OB), inner bark (IB), sapwood (SW), and heartwood (HW) at different vertical heights of *Q. serrata*. Samples collected from 1/2 to 3/4 tree height were combined together as the upper parts, and samples collected from 0.2 m were called the lower parts. Error bars indicate the standard deviations of the three replicates in the lower parts and the 3–6 replicates in the upper parts. The y-axis of 1.2 m is in logarithmic units

4.2.4 *^{133}Cs in Bark, Wood, and Needles of* P. densiflora

The concentration of ^{133}Cs in different components of *P. densiflora* was given in Table 4.4. At 0.2, 1.2, and 2.2 m, the ^{133}Cs concentrations in the outer bark, inner bark, sapwood, and heartwood were much higher in ^{133}Cs-treated trees than control trees, although there were no significant differences among them. The highest ^{133}Cs concentration in sapwood at 1.2 m, the height at which the $^{133}CsCl$ solution was applied, suggested that ^{133}Cs penetrated the wood tissue through the bark of *P. densiflora*. Unfortunately, because the soil under treated trees was contaminated

Table 4.4 ^{133}Cs concentrations in the outer bark, inner bark, sapwood, heartwood, needles, and branches of *P. densiflora*

Position	Components	Cs (μg g^{-1})	
		Treatment	Control
Tree top	Needles	0.19 ± 0.10	0.13 ± 0.021
	Branches	0.045 ± 0.011	0.11 ± 0.015
2.2 m	Outer bark	0.067 ± 0.013	0.039 ± 0.014
	Inner bark	0.071 ± 0.0067	0.028 ± 0.016
	Sapwood 1	0.015 ± 0.0018	0.0030 ± 0.0012
	Sapwood 2	0.012 ± 0.0019	n.d.
	Heartwood	n.d.	n.d.
1.2 m	Outer bark	9.1 ± 2.0	0.018 ± 0.0035
	Inner bark	8.3 ± 1.2	0.019 ± 0.010
	Sapwood 1	0.10 ± 0.010	0.0017 ± 0.00023
	Sapwood 2	0.027 ± 0.0088	n.d.
	Heartwood	0.0083 ± 0.0030	n.d.
0.2 m	Outer bark	16 ± 2.9	0.014 ± 0.0021
	Inner bark	5.6 ± 0.23	0.011 ± 0.0038
	Sapwood 1	0.11 ± 0.062	0.0040 ± 0.0011
	Sapwood 2	0.050 ± 0.026	n.d.
	Heartwood	0.0065 ± 0.0033	n.d.

Concentrations are average ± SE of three replicates
Disks samples were collected from 0.2 m, 1.2 m, to 2.2 m
n.d. not detected

with the applied solution, the possible contribution of root uptake to the loading of ^{133}Cs in the wood cannot be denied. The higher ^{133}Cs concentration in the outer bark at 0.2 m also implied that blocking the stemflow by urethane might not be effective (Table 4.4). One possible reason is that during our experiment period, the intermittent rain days and several heavy rain days (e.g., 87 mm on 10 September) may lead to serious leakage of ^{133}Cs.

4.3 Absorption and Translocation of Cs Through the Bark

Due to the absence of ^{133}Cs contamination in the soils around the ^{133}Cs-treated trees, the highest ^{133}Cs concentration in the sapwood at 1.2 m, where the ^{133}CsCl solution was applied, demonstrates that ^{133}Cs in the surface of the bark penetrated into the wood tissue through the bark of all the three tree species.

The mechanism for the Cs absorption through the bark has not yet been clearly identified. Several studies reported the absorption of trace metals through the bark; however, the mechanisms by which these trace metals enter the wood through the bark were also unclear (Zhang et al. 1995; Watmough and Hutchinson 2003). The bark consists of two layers, the inner bark and outer bark that are separated by a moisture-impervious layer (Watmough and Hutchinson 2003). The outer bark (rhytidome) is composed entirely of dead tissue that consists of remaining cell walls composed mainly of lignin, cellulose, hemicellulose, pectin, and suberin (Rulik et al. 2014). The inner bark consists of remaining living tissues comprising phloem and the innermost phellogen and phelloderm (Borger 1973). The bark has frequently been used as an effective sorbent for the removal of metal ions from wastewater (Aoyama et al. 2004; Argun et al. 2009). The sorption of metal ions to the bark takes place via ion exchange, primarily via complexation to functional groups, such as carboxyl and phenolic hydroxyl groups (Su et al. 2013). Takenaka and Sasama (2000) suggested that an ion exchange occurs between cations contained in the stemflow and protons in the stem. We speculate that the absorption of Cs by the outer bark is related to an ion-exchange process where Cs is exchanged with other cations, such as protons. The infiltration of Cs from the outer bark to the inner bark is likely a non-metabolic process because the outer bark consists of fully dead tissues formed from cork cambium (Watmough and Hutchinson 2003). However, the mechanism by which Cs enters the wood through the inner bark is still unclear. Lepp and Dollard (1974) demonstrated the lateral movement of lead from the bark to the wood and stated that cambial activity is not essential for this process to take place, while other studies have reported that the transfer of solutions from the inner bark (phloem) to the sapwood (xylem) is predominantly facilitated through the symplast in uniseriate ray parenchyma cells and driven by transpiration and associated gradients between the phloem and xylem (Pfautsch et al. 2015). The penetration of applied ^{133}Cs into the wood in dormant season (from 28 September 2014 to 23 November 2014) of *Q. serrata* in our study indicated that the absorption process was probably independent of cambial activities, i.e., Cs could be absorbed into the wood through the bark throughout the year if Cs exists in a soluble form. Further studies of the interaction between the phloem and xylem (Hölttä et al. 2009) are required to fully understand the transfer of Cs from the inner bark to the wood.

4.4 Distribution of Cs Within Trees

In the case of *C. japonica*, to clarify the radial distribution pattern of bark-derived ^{133}Cs in the wood, the concentration ratio of ^{133}Cs in the heartwood to sapwood of 1.2-m disks taken from ^{133}Cs-treated trees and control trees at Utsunomiya site was calculated (Fig. 4.4). The ratio tended to be higher in the control trees than the treated trees, and ^{133}Cs tended to accumulate in the heartwood (the ratio>1). This observation agrees with previous findings for ^{137}Cs distribution in *C. japonica* (Chigira et al. 1988; Masuchika et al. 1988; Momoshima et al. 1995). In the same figure, the ratios of K, which is the chemical analogue of Cs, were plotted, and the ratios were around 2.5–3.0. The accumulation of alkali metals in the heartwood of *C. japonica* might be associated with heartwood formation (Momoshima et al. 1995; Kubo and Ataka 1998; Kuroda et al. 2013), pH regulation (Morikawa et al. 1996), or promotion of polyphenol dissociation (Okada et al. 2012). Thus, it was considered that cesium

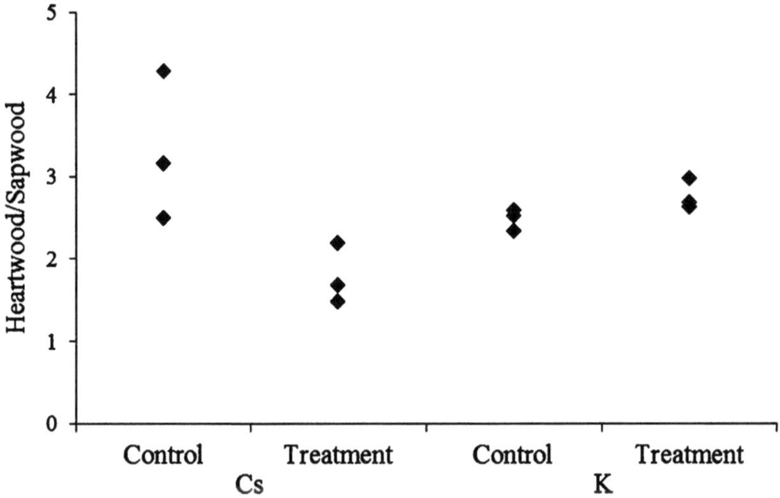

Fig. 4.4 The concentration ratios of heartwood to sapwood for ^{133}Cs and K at 1.2-m disks for control trees and ^{133}Cs-treated trees at Utsunomiya site ($n = 3$)

absorbed by bark was translocated toward the heartwood, probably together with K, via ray parenchyma cells, though the mechanism for such movement is still unclear. In addition, the ratios derived from root uptake in the control trees tended to be higher than the ratios affected by surface absorption through the bark in the treated trees. This means that the treatment period in the present experiment (2.5 months) might not be enough to attain the steady state of ^{133}Cs distribution in the disks of *C. japonica*. Higher ^{133}Cs concentrations in current-year needles of ^{133}Cs-treated trees than controls trees suggested that ^{133}Cs absorbed by bark was also translocated vertically to apex (Tables 4.2 and 4.3), comprising assimilating organs (new needles) and physiologically active tissue (1-year-old shoots) (Ipatyev et al. 1999). In addition, ^{133}Cs concentrations in the needles of 10-year-old trees were higher than in 26-year-old trees, which might have been caused by biomass dilution, whereby older trees have higher leaf biomass than younger trees. In addition, such difference in ^{133}Cs concentration could be partly due to seasonal variations of radiocesium in leaves. The sampling time at Utsunomiya was in October; radiocesium might be translocated from leaves to other tree organs such as trunk and twigs in autumn–winter (dormancy period) (Sombré et al. 1994; Yoshihara et al. 2014).

The radial distribution of ^{133}Cs in *Q. serrata* shows some difference from *C. japonica* (Fig. 4.3), at a height of 1.2 m; the applied ^{133}Cs was translocated only into the sapwood but not into heartwood over 2 months. Ohashi et al. (2014) also reported the similar results of abrupt decrease of cesium migration to heartwood of *Q. serrata* after FDNPP accident. The abrupt decrease of Cs concentration in heartwood indicated that some barrier may be present at the sapwood–heartwood boundaries. The comparable ^{133}Cs concentration between the ^{133}Cs-treated trees and the control trees in the wood above and below 1.2 m, as well as the leaves and branches, suggested that vertical translocation of Cs was negligible, probably due to that during our experimental periods, tree growth and leaf photosynthetic activity may significantly drop, and thus transpiration flow also become very small.

For *P. densiflora*, higher ^{133}Cs concentration in heartwood of treated trees than control trees suggested that the applied ^{133}Cs was translocated until heartwood (Table 4.4). No evident difference was observed for ^{133}Cs concentration in needles and branches between ^{133}Cs-treated trees and control trees, which suggested that ^{133}Cs was not translocated into needles during our experimental periods.

The different radial distribution patterns of radiocesium in different tree species (Kohno et al. 1988; Thiry et al. 2002) can be explained by the difference in their radial ray compositions (Soukhova et al. 2003), by which Cs is translocated through the sapwood to the heartwood. Because no living cells exist in the heartwood of all the tree species, other contributing factors in addition to radial ray compositions, such as water content and the role of heartwood formation, require further investigation to fully understand the translocation and distribution of Cs in the wood of different tree species.

Absorption of Cs into the bark and then translocation to the inner tissues of trees might play a significant role in tree contamination, especially in the early stages of ^{137}Cs deposition when root uptake might be low. This finding is of particular importance for deciduous trees because no leaves were present at the time of the FDNPP accident; hence, much more ^{137}Cs was potentially deposited directly onto the bark surface. The ^{137}Cs absorbed on the bark probably exists in a stable chemical form and is not easily washed off the bark (Kato et al. 2012; Iwase et al. 2013). No evident changes in ^{137}Cs concentrations in the bark of *C. japonica* were observed 2 years after the Fukushima accident (Kajimoto et al. 2015), and the ^{137}Cs concentrations in the bark of *Quercus petraea* remained high even 22 years after the Chernobyl accident (Zhiyanski et al. 2010). However, if the stable form of Cs in the bark degrades and releases ^{137}Cs ions, then absorption and translocation of ^{137}Cs ions from the bark into the tree might occur continuously, hence might become an important source of pollution for forest trees in the coming years.

4.5 Conclusion

In this study, we confirmed the absorption of Cs through the bark of three tree species under field conditions. The bark-derived Cs exhibited different radial distribution patterns among tree species. Cs tends to accumulate in the heartwood of *C. japonica*, while Cs concentration decreased toward heartwood of *Q. serrata* and *P. densiflora*. Cs was translocated to tree top of *C. japonica*, but this phenomenon was not clear for *Q. serrata* and *P. densiflora*. Absorption of Cs through the bark to the sapwood might be a passive diffusion process and might depend on bark thickness and bark structure which need further research.

4.6 Materials and Methods

4.6.1 *Application of ^{133}Cs to Bark of Three Tree Species*

A photograph of the ^{133}Cs application process to a representative *C. japonica* in the field is shown in Fig. 4.5. Initially, a 40-ml aliquot of 0.01 M $^{133}CsCl$ solution was spread uniformly on a paper towel. The towel was then attached to the bark at 1.2 m above the ground. To prevent the towel from drying, it was covered on the outside with plastic wrap and fixed using plastic wires. Furthermore, to prevent soil contamination from $^{133}CsCl$ caused by stemflow, a 3-cm thick urethane sheet was wrapped around the trunk below the paper towel, which was then fixed using plastic wires that also functioned as stemflow gutters. A hose was connected to the urethane sheet to direct the stemflow to the gutterway or a bucket. The juncture between hose, bark, and urethane was filled with silicon resin to prevent leakage. Both treatments and controls had three replicates at each site. The experiment sites, average tree age, and experimental periods for each tree species are shown in Table 4.5.

4.6.2 *Sampling, Pretreatment, and Chemical Analysis*

After experiment, 10-cm-thick disks were collected from several different heights of each tree species (Table 4.5); current-year leaves/needles were sampled from the trees' canopies; the surface soil (0–5 cm) was also collected around each tree to

Fig. 4.5 A photograph of the ^{133}Cs application experiment on a representative 10-year-old cedar. A paper towel with 40 mL 0.01 M $^{133}CsCl$ solution was wrapped around bark at 1.2 m. A urethane sheet and hose were used to direct the stemflow to guttering

Table 4.5 Information of study sites, trees, experimental periods, and sampling height

Tree species	Sites	Average tree age	Experimental periods	Sampling height (m)
C. japonica	Koriyama (37°21′N, 140°20′E)	10	31/07/2013–22/08/2013	0.2
				1.2
				2.2
	Utsunomiya (36°46′N,139°49′E)	26	06/08/2013–22/10/2013	0.2
				1.2
				6.2
Q. serrata	Yamakiya (37° 45.5′ N, 140° 28.2′ E)	50	28/09/2014–23/11/2014	0.2
				1.2
				1/2 H
				3/4 H
P. densiflora	Setohachiyama (37°45.5′N, 140°28.2′E)	42	02/06/2015–25/09/2015	0.2
				1.2
				2.2
				1/2 H

1/2 H and 3/4 H mean 1/2 the tree height (1/2 H) and 3/4 the tree height, respectively

assess whether the soil was contaminated with ^{133}Cs (n = 3 for each). For disk samples, in the case of *C. japonica*, the disks were separated into the bark, sapwood 4 mm away from the cambium, and heartwood. For *Q. serrata* and *P. densiflora*, each disk was divided into outer bark, inner bark, sapwood, and heartwood, the sapwood was separated into two to three segments according to their distances from the cambium, and the average distance of each segment was sapwood 1 (SW1): 0–0.5 cm, sapwood 2 (SW2): 0.5–1.5 cm, and sapwood 3 (SW3): 1.5–3 cm. The bark samples were ultrasonically washed three times with ultrapure water for 10 s to remove contaminants and non-absorbed ^{133}Cs. The wood samples were not washed, but the contaminated surfaces were removed using a hammer and stainless steel chisel. The needles/leaves were washed with tap water and then rinsed with ultrapure water. All plant samples were oven dried at 80 °C for more than 48 h to a constant weight and then homogenized to a powder. Soil samples were air dried for more than 1 month, passed through a 2-mm sieve, and stored for analysis.

The plant samples (0.1 g) were digested with HNO_3 using a graphite block acid digestion system (EcoPre; ODLAB). The soil samples (2 g) were extracted using 1 M of NH_4AC (40 ml, pH 7) to determine the ion-exchangeable ^{133}Cs. The concentration of ^{133}Cs was determined using an inductively coupled plasma mass spectrometer (iCAPQc, Thermo Scientific) with In as the internal standard. The concentration of K was analyzed via inductively coupled plasma atomic emission spectrometry (ICP-AES; IRIS ICARP, Jarrell Ash Nippon Corp., Kyoto, Japan).

References

Aliyu AS, Evangeliou N, Mousseau TA, Wu J, Ramli AT (2015) An overview of current knowledge concerning the health and environmental consequences of the Fukushima Daiichi Nuclear Power Plant (FDNPP) accident. Environ Int 85:213–228

Aoyama M, Kishino M, Jo TS (2004) Biosorption of Cs (VI) on Japanese cedar bark. Sep Sci Technol 39:1149–1162

Argun ME, Dursun S, Karatas M (2009) Removal of Cd (II), Pb (II), Cu (II) and Ni (II) from water using modified pine bark. Desalination 249:519–527

Borger GA (1973) Development and shedding of bark. In: Kozlowski TT (ed) Shedding of plant parts. Academic, New York, pp 205–236

Calmon P, Thiry Y, Zibold G, Rantavaara A, Fesenko S (2009) Transfer parameter values in temperate forest ecosystems: a review. J Environ Radioact 100:757–766

Chigira M, Saito Y, Kimura K (1988) Distribution of ^{90}Sr and ^{137}Cs in annual rings of Japanese cedar, *Cryptomeria japonica* D. Don. J Radiat Res 29:152–160

Endo I, Ohte N, Iseda K, Tanoi K, Hirose A, Kobayashi NI, Murakami M, Tokuchi N, Ohashi M (2015) Estimation of radioactive 137-cesium transportation by litterfall, stemflow and throughfall in the forests of Fukushima. J Environ Radioact 149:176–185

Hasegawa M, Ito MT, Kaneko S, Kiyono Y, Ikeda S, Makino SI (2013) Radiocesium concentrations in epigenic earthworms at various distances from the Fukushima nuclear power plant 6 months after the 2011 accident. J Environ Radioact 126:8–13

Hashimoto S, Ugawa S, Nanko K, Shichi K (2012) The total amounts of radioactively contaminated materials in forests in Fukushima, Japan. Sci Rep 2:416

Hölttä T, Mencuccini M, Nikinmaa E (2009) Linking phloem function to structure: analysis with a coupled xylem-phloem transport model. J Theor Biol 259:325–327

Ipatyev V, Bulavik I, Baginsky V, Gonchrenko G, Dvornik A (1999) Forest and Chernoyl: forest ecosystems after the Chernobyl nuclear power plant accident: 1986–1994. J Environ Radioact 42:9–38

Iwase K, Tomioka R, Sugiura Y, Kanasashi T, Takenaka C (2013) Absorption properties of the Cs in the bark of *Cryptomeria japonica* and *Quercus serrata*. Jpn J For Environ 55:69–73. (in Japanese)

Kajimoto T, Saito S, Kawasaki T, Kabeya D, Yazaki K, Tanaka H, Ota T, Matsumoto Y, Tabuchi R, Kiyono Y, Takano T, Kuroda K, Fujiwara T, Suzuki Y, Komatsu Ohashi S, Kaneko S, Akama A, Takahashi M (2015) Dynamics of radiocesium in forest ecosystems affected by the Fukushima Daiichi Nuclear Power Plant accident: species-related transfer processes of radiocesium from tree crowns to ground floor during the first two years. J Jpn For Soc 97:33–43. (in Japanese)

Kato H, Onda Y, Gomi T (2012) Interception of the Fukushima reactor accident-derived ^{137}Cs, ^{134}Cs and ^{131}I by coniferous forest canopies. Geophys Res Lett 39:L20403

Kohno M, Koizumi Y, Okumura K, Mito I (1988) Distribution of environmental Cesium-137 in tree rings. J Environ Radioact 8:15–19

Kubo T, Ataka S (1998) Blackening of sugi (*Cryptomeria japonica* D. Don) heartwood in relation to meta content and moisture content. J Wood Sci 44:137–141

Kuroda K, Kagawa K, Tonosaki M (2013) Radiocesium concentrations in the bark, sapwood and heartwood of three tree species collected at Fukushima forests half a year after the Fukushima Dai-ichi nuclear accident. J Environ Radioact 122:37–42

Lepp NW, Dollard GJ (1974) Studies on lateral movement of 210Pb in woody stems. Oecologia 16:179–184

Mahara Y, Ohta T, Ogawa H, Kumata A (2014) Atmospheric direct uptake and long-term fate of radiocesium in trees after the Fukushima nuclear accident. Sci Rep 4:7121

Masuchika K, Yoshinobu K, Katsuno O, Iwako M (1988) Distribution of environmental Cesium-137 in tree rings. J Environ Radioact 8:15–19

Momoshima N, Eto I, Kofuji H, Takashima Y, Koike M, Imaizumi Y (1995) Distribution and chemical characteristics of cations in annual rings of Japanese cedar. J Environ Qual 24:1141–1149

Morikawa T, Oda K, Matsumura J, Oda K, Utsumi Y, Yamamoto K (1996) Black-heartwood formation and ash contents in the stem of sugi (*Cryptomeria japonica D. Don*) II: heart-wood properties of three sugi cultivars (in Japanese with English summary). Bull Kyushu Univ For 74:41–49

Nishikiori T, Watanabe M, Koshikawa MK, Takamatsu T, Ishii Y, Ito S, Takenaka A, Watanabe K, Hayshi S (2015) Uptake and translocation of radiocesium in cedar leaves following the Fukushima nuclear accident. Sci Total Environ 52:611–616

Ohashi S, Okad N, Tanaka A, Nakai W, Takano S (2014) Radial and vertical distributions of radiocesium in tree stems of *Pinus densiflora* and *Quercus serrata* 1.5 y after the Fukushima nuclear disaster. J Environ Radioact 134:54–60

Ohta T, Torimoto J, Kubota T, Mahara Y (2016) Front tracking of the translocation of water-soluble cesium deposited on tree leaves of plum. J Radioanal Nucl Chem 310:109–115

Okada N, Hirakawa Y, Katayama Y (2012) Radial movement of sapwood-injected rubidium into heartwood of Japanese cedar (*Cryptomeria japonica*) in the growing period. J Wood Sci 58:1–8

Pfautsch S, Renard J, Tjoelker MG, Salih A (2015) Phloem as capacitor: radial transfer of water into xylem of tree stems occurs via symplastic transport in ray parenchyma. Plant Physiol 167:963–971

Rulik P, Pilatova H, Suchara I, Sucharova J (2014) Long-term behaviour of Cs-137 in spruce bark in coniferous forests in the Czech Republic. Environ Pollut 184:511–514

Sato M, Takata D, Tanoi K, Ohtsuki T, Muramatsu Y (2015) Radiocesium transfer into the fruit of deciduous fruit trees contaminated during dormancy. Soil Sci Plant Nutr 61:156–164

Sombré L, Vanhouche M, de Brouwer S, Ronneau C, Lambotte JM, Myttenaere C (1994) Long-term radiocesium behavior in spruce and oak forests. Sci Total Environ 157:59–71

Soukhova NV, Fesenko SV, Klein D, Spiridonov SI, Sanzharova NI, Badot PM (2003) ^{137}Cs distribution among annual rings of different tree species contaminated after the Chernobyl accident. J Environ Radioact 65:19–28

Su PP, Granholm K, Pranovich A, Harju L, Holmbom B, Ivaska A (2013) A sorption of metal ions from aqueous solution to spruce bark. Wood Sci Technol 47:1083–1097

Sugiura Y, Shibata M, Ogata Y, Ozawa H, Kanasashi T, Takenaka C (2016) Evaluation of radiocesium concentrations in new leaves of wild plants two years after the Fukushima Dai-ichi Nuclear Power Plant accident. J Environ Radioact 160:8–24

Tagami K, Uchida S, Ishii N, Kagiya S (2012) Translocation of radiocesium from stems and leaves of plants and the effect on radiocesium concentration in newly emerged plant tissues. J Environ Radioact 111:65–69

Takenaka C, Sasama T (2000) Ion exchange reactions on the stem surface of *Chamaecyparis obtusa* Sieb. et Zucc. Trees 14:354–360

Thiry Y, Goor F, Riesen T (2002) The true distribution and accumulation of radiocaesium in stem of Scots pine (*Pinus sylvestris* L.,). J Environ Radioact 58:243–259

Watmough SA, Hutchinson TC (2003) Uptake of ^{207}Pb and ^{111}Cd through bark of mature sugar maple, white ash and white pine: a field experiment. Environ Pollut 121:39–48

Yoshida S, Muramatsu Y, Dvornik AM, Zhuchenko TA, Linkov I (2004) Equilibrium of radiocesium with stable cesium within the biological cycle of contaminated forest ecosystems. J Environ Radioact 75:301–313

Yoshihara T, Matsumura H, Kobayashi T, Hashida SN, Nagaoka T, Tsuzaki M, Wakamatsu T, Goto F (2014) Changes in radiocesium contamination from Fukushima in foliar parts of 10 common tree species in Japan between 2011 and 2013. J Environ Radioact 138:220–226

Zhang L, Qian JL, Planas D (1995) Mercury concentration in tree rings of black spruce (*Picea mariana* Mill. B.S.P.) in boreal Quebec, Canada. Water Air Soil Pollut 81:163–173

Zhiyanski M, Sokolovska M, Bech J, Clouvas A, Penev I, Badlin V (2010) Cesium-137 contamination of oak (*Quercus petraea* Liebl.) from sub-Mediterranean zone in South Bulgaria. J Environ Radioact 101:864–886

Chapter 5
Translocation of ^{137}Cs in the Woody Parts of Sugi (*Cryptomeria japonica*)

Kazuya Iizuka, Jyunichi Ohshima, Futoshi Ishiguri, Naoko Miyamoto, Mineaki Aizawa, Tatsuhiro Ohkubo, Chisato Takenaka, and Shinso Yokota

Abstract To clarify the behavior of ^{137}Cs absorbed in the stem wood 5 years after the Fukushima Daiichi Nuclear Power Plant accident, ^{137}Cs concentration was measured in a radial direction, from sapwood to heartwood, in the stem wood of 33-year-old sugi (*Cryptomeria japonica*) trees. Understanding the mechanism of absorption and movement of ^{137}Cs to stem wood is necessary while searching for clues that may help to predict the transit of ^{137}Cs into the xylem tissues. Therefore, we investigated the radial variation in ^{137}Cs concentration in sugi, trees, and then we tried to elucidate the relationship between ^{137}Cs concentration and potassium (K) content. A group of trees with a high heartwood:sapwood ^{137}Cs concentration ratio showed significantly higher K and moisture contents and lower lightness value (L^*) than those of another group of trees, which showed lower K concentrations in the wood. Heartwood:sapwood ^{137}Cs concentration ratio was significantly and positively correlated with heartwood:sapwood K content ratio. Based on these results, we propose that the translocation of ^{137}Cs from sapwood to heartwood is related to the heartwood:sapwood K content ratio in sugi trees.

Keywords ^{137}Cs concentration · Potassium content · Radial direction · Stem wood · Sugi

---

K. Iizuka (✉)
School of Agriculture, University Forests, Utsunomiya University, Shioya-gun, Tochigi, Japan
e-mail: kiizuka@cc.utsunomiya-u.ac.jp

J. Ohshima · F. Ishiguri · M. Aizawa · T. Ohkubo · S. Yokota
School of Agriculture, Utsunomiya University, Utsunomiya, Tochigi, Japan

N. Miyamoto
Tohoku Regional Breeding Office, Forest Tree Breeding Center, Forestry and Forest Products Research Institute, Ohsaki Takizawa, Iwate, Japan

C. Takenaka
Graduate School of Bioagricultural Sciences, Nogoya University, Nagoya, Aichi, Japan

C. Takenaka et al. (eds.), *Radiocesium Dynamics in a Japanese Forest Ecosystem*,
https://doi.org/10.1007/978-981-13-8606-0_5

5.1 Introduction

A large amount of radioactive material was released into the environment after the accident at the Fukushima Daiichi Nuclear Power Plant caused by the Great East Japan Earthquake that occurred in March 2011. Subsequently, the radioactive fallout contaminated a vast forest area that is important to forestry and forest industry and a major contributor to the regional industry. Studies on radioactive fallout from atmospheric nuclear tests have shown that ^{137}Cs, which has a relatively long half-life (30.2 years), is characterized by high mobility within the stem wood (Kohno et al. 1988; Kudo et al. 1993; Momoshima and Bondietti 1994; Kagawa et al. 2002). Approximately 90% of the ^{137}Cs released from the nuclear plant still remains, 5 years after the accident. Thus, environmental contamination by ^{137}Cs deposition is still present in the forests in the Fukushima and neighboring prefectures. Several studies conducted during the early stages after the Fukushima accident showed the effects of radioactive fallout absorbed by trees, especially the concentration of radioactive cesium in the xylem (Kuroda et al. 2013; Ohashi et al. 2014, 2017; Mahara et al. 2014; Ogawa et al. 2016; Nagakura et al. 2016; Imamura et al. 2017). The tree species investigated here were sugi, also known as Japanese red cedar (*Cryptomeria japonica*), and akamatsu (*Pinus densiflora*) as examples of evergreen coniferous trees and konara (*Quercus serrata*) as an example of a deciduous broad-leaved tree. The investigations in the early stages after the accident detected radiocesium in the woody parts of these tree species, which is another reason why these trees were selected for our study.

Zhu and Smolders (2000) reported that radiocesium uptake occurs mainly by two transport pathways in root cell membranes, namely, the K^+ transporter and K^+ channel pathway. Studies of contaminated trees revealed uptake and migration of ^{137}Cs via the leaves (Tagami et al. 2012; Nishikiori et al. 2015). On the other hand, in the case of sugi, Wang et al. (2016) proved that cesium was absorbed by the stem wood through the bark. Further, the movement of ^{137}Cs from sapwood to heartwood in the stem wood of sugi tree might be related to the heartwood:sapwood K content ratio (Iizuka et al. 2018). Nagakura et al. (2016) elucidated the relation between ^{137}Cs concentration, ^{133}Cs concentration, potassium content, and rubidium content in stem wood and leaves of sugi trees in testing plots, about 3.5 years after the accident. In addition, Imamura et al. (2017) reported temporal changes in the distribution of radioactive cesium in forests 5 years after the accident. In some testing plots located in Fukushima Prefecture, ^{137}Cs concentration increased significantly from 2011 to 2015 in the wood of sugi, hinoki (*Chamaecyparis obtusa*), and konara trees. Ohashi et al. (2017) reported a similar trend in the above-mentioned test sites in 2016.

Trends in future change of ^{137}Cs concentration in the wood are of particular interest to the forestry and wood industry. Sugi trees occupy the largest planting area in Japan, and thus, Sugi has the largest influence on the timber industry. Several studies have explored the level of ^{137}Cs absorbed in stem wood of sugi trees early after the accident (Kuroda et al. 2013; Mahara et al. 2014; Ogawa et al. 2016; Nagakura et al. 2016; Wang et al. 2016; Iizuka and Ohshima 2017; Imamura et al.

2017; Ohashi et al. 2017; Iizuka et al. 2018); ^{137}Cs concentration was found to be higher in the heartwood than in the sapwood in the first 1–2 years after the accident (Iizuka and Ohshima 2017; Ohashi et al. 2017). It is important to clarify the within-tree distribution of ^{137}Cs based on a large number of samples from sugi trees. In addition, the movement of ^{137}Cs in the stem wood should be determined in order to predict the future level of contamination of sugi plantation forests and wood resources from plantations.

Potassium (K), an alkali metal in the same group as cesium, is closely related to the blackening phenomenon in the heartwood of sugi trees. Heartwood with higher K contents have a higher moisture content and a lower lightness (L^*) value in the $L^*a^*b^*$ system (Kawazumi et al. 1991; Abe et al. 1994; Oda et al. 1994; Morikawa et al. 1996; Kubo and Ataka 1998; Ishiguri et al. 2006; Miyamoto et al. 2016). Okada et al. (1987) reported that K and other alkali metal contents in Japanese cedar showed an abrupt increase from sapwood to heartwood and almost constant values in the heartwood. They also found that K and other alkali metals are actively transported from the sapwood to the outer heartwood via the rays, resulting in their diffusion and accumulation in the heartwood of Japanese cedar (Okada et al. 2011, 2012). Furthermore, several studies have investigated the relationship between ^{137}Cs and ^{40}K in Japanese cedar wood (Katayama et al. 1986; Chigira et al. 1988; Kudo et al. 1993). However, information on the variation of ^{137}Cs concentration in the wood of trees exposed to the radioactive fallout from the Fukushima accident is scarce. Further, detailed information on the variation of radiocesium concentration within a tree stem is scarce. Therefore, further research is required on these relationships at an early stage after radioactive fallout, by using a large number of trees. In addition, in order to clarify the mechanism of radiocesium transfer in the stem of sugi trees, it is also necessary to document K, moisture content, and lightness (L^* value) in sugi heartwood.

In the present study, ^{137}Cs concentration, K content, moisture level, and lightness (L^*) of heartwood were investigated in sugi trees grown in a plantation located 130 km southwest of the Fukushima Daiichi Nuclear Power Plant. The movement of ^{137}Cs from sapwood to heartwood and its relationship to K content were investigated. Our data aim to provide information that will help to predict future behavior of ^{137}Cs in the stem wood of forest species.

5.2 Cesium-137 Concentration in Surface Mineral Soils and Stem Wood of Sugi Trees Sampled in the Surveyed Area

Mean ^{137}Cs concentration in the surface mineral soils (topsoil) at the 18 sampled points was 1727 ± 1130 (SD) Bq kg^{-1}; the high coefficient of variation (CV = 65%) measured showed a large variation in ^{137}Cs concentration over the surveyed area 5 years after the accident. Therefore, ^{137}Cs contamination of the soil surface was quite heterogeneous in the surveyed area.

Table 5.1 ^{137}Cs concentration in the bark, sapwood (SW), and heartwood (HW) and the correlation coefficient of each indicator

	^{137}Cs (Bq kg^{-1})				
Indicator	Mean	Std.	Code	X1	X2
Bark	209	118	X1	1.000	
SW	20	11	X2	0.411**	1.000
HW	43	26	X3	0.196 ns	0.758**

std. standard deviation, ns no significance
**Significant at 1% level

The ^{137}Cs concentrations in the bark, sapwood, and heartwood obtained from 80 individuals at 0.2 m above the ground and the correlation coefficients of each indicator are shown in Table 5.1. The mean diameter of the wood disk samples was 19.7 cm, and the heartwood occupied 39.2% of the disk area. Mean ^{137}Cs concentrations in the bark, sapwood, and heartwood were 209, 21, and 43 Bq kg^{-1}, respectively. A significant and positive correlation was found for ^{137}Cs concentration between bark and sapwood ($r = 0.411^{**}$) and between sapwood and heartwood ($r = 758^{**}$).

It has been assumed that, from the total amount of ^{137}Cs deposited on the bark by radioactive fallout at the beginning of the accident, soluble ^{137}Cs has been absorbed into the xylem, whereas the contamination detected in the bark (outer bark + inner bark) would come from the ^{137}Cs deposited on the outer bark directly from the atmosphere. Therefore, 5 years after the accident, the remaining ^{137}Cs in the bark may be fixed/absorbed physically or chemically on the bark. Iwase et al. (2013) reported that Cs fixed/absorbed in the tree bark of Japanese cedar is not easily removed. In contrast, ^{137}Cs in the xylem is thought to be present in a soluble form and is therefore mobile. ^{137}Cs fixed/absorbed into the bark and that absorbed in the xylem may represent two different chemical forms of ^{137}Cs. These facts indicate that 5 years after the accident, the bark and xylem contain two different chemical forms of ^{137}Cs, with the one in the bark being immobile. Meanwhile, ^{137}Cs in the xylem, which is composed of sapwood and heartwood, is found in a soluble and mobile chemical form, which explains the relatively high correlation coefficient ($r = 0.758^{**}$) for ^{137}Cs concentration in sapwood as it relates to that in the heartwood.

5.3 Radial Movement of ^{137}Cs in the Stem Wood

The mechanism whereby ^{137}Cs moved from sapwood to heartwood was examined. The distribution of ^{137}Cs concentration in sapwood and heartwood is shown in Fig. 5.1. The range of ^{137}Cs concentration in the heartwood was wider than that in sapwood. The movement velocity of ^{137}Cs from sapwood to heartwood may be related to the heartwood:sapwood ^{137}Cs concentration ratio (Fig. 5.2). Mean

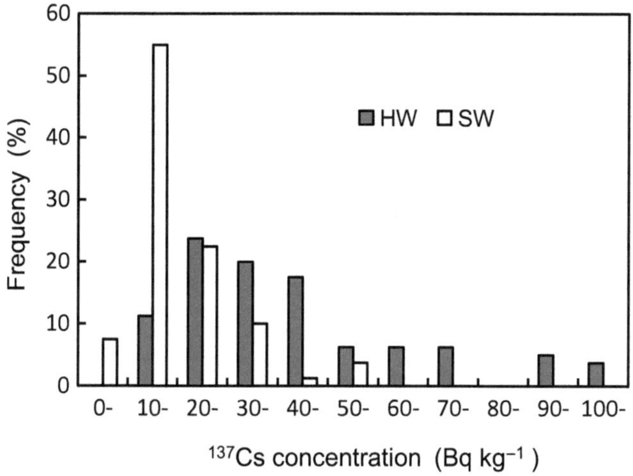

Fig. 5.1 Frequency of ^{137}Cs concentration in heartwood (HW) and sapwood (SW)

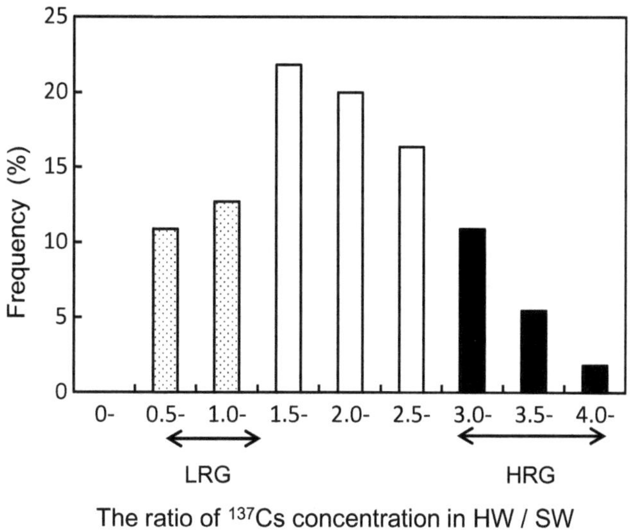

Fig. 5.2 Frequency of the ratio of ^{137}Cs concentration in heartwood (HW) to sapwood (SW)

heartwood:sapwood ^{137}Cs concentration ratio was 2.1 ± 0.9 (SD). Based on this mean and standard deviation values, two groups of ten individuals were selected, such that one group (HRG) was composed of individuals with the heartwood: sapwood ^{137}Cs concentration ratio $\geqq 3.0$, while the other group (LRG) comprised those with a heartwood:sapwood ^{137}Cs concentration ratio $\leqq 1.4$; the underlying assumption being that the movement velocity of ^{137}Cs from sapwood to heartwood would be faster in the former group.

Table 5.2 ^{137}Cs concentration, potassium content, moisture content, and L^* value in two categories of sampling trees and the result in *t*-test

			^{137}Cs concentration (Bq kg^{-1})			Potassium content (g kg^{-1})			MC (%)	L^*
Category	n		HW/SW	HW	SW	HW/SW	HW	SW	HW	HW
HRG	10	Mean	3.4	68	21	4.9	4.0	0.8	107	64.77
		Std.	0.5	38	11	0.8	1.0	0.3	2.9	2.77
LRG	10	Mean	0.9	22	23	1.9	1.5	0.8	69	71.53
		Std.	0.3	13	8	0.8	0.8	0.4	16	1.73
t-test			*	*	ns	*	*	ns	*	*

MC Moisture Content, *std.* standard deviation, *ns* no significance
*Significant at 5% level

Table 5.3 Correlation coefficient between each indicator

Indicator		HW/SW ^{137}Cs	K
HW/SW	^{137}Cs	1.000	
	K	0.920**	1.000
Heartwood	K	0.823**	0.779**
	MC	0.666**	0.572**
	L^*	−0.865**	−0.792**

**Significant at 1% level

Cesium-137 concentration, K content, moisture content, and L^* value in HRG and LRG are shown in Table 5.2. ^{137}Cs concentration and K content in heartwood was significantly higher in HRG than in LRG, but no statistical difference was observed in sapwood between the two groups. A significant difference in moisture content and L^* value in heartwood was found between the two groups; thus, HRG showed a higher moisture content and lower L^* value as compared with LRG. K content in heartwood was positively correlated with moisture content but negatively correlated with L^* (Kawazumi et al. 1991; Abe et al. 1994; Oda et al. 1994; Morikawa et al. 1996; Kubo and Ataka 1998; Ishiguri et al. 2006; Miyamoto et al. 2016). Furthermore, it has been proposed that these indicators and their relationships may be genetically regulated (Miyamoto et al. 2016). Abe et al. (1994) identified potassium bicarbonate as a causative agent of heartwood color.

The correlation coefficients between indicator variables are shown in Table 5.3. A highly significant and positive correlation was observed between the heartwood: sapwood ^{137}Cs concentration ratio and the corresponding K content ratio (Fig. 5.3, $r = 0.920^{**}$). A significant correlation was found between heartwood:sapwood ^{137}Cs concentration ratio and K content, moisture content, and L^* value in heartwood; similarly, a significant correlation was found between heartwood:sapwood K content ratio and K content, moisture content, and L^* value in heartwood. The movement of ^{137}Cs from sapwood to heartwood may be also affected by moisture content in the heartwood.

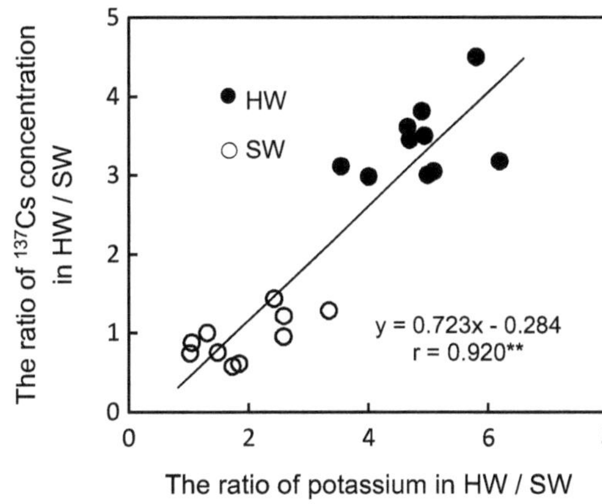

Fig. 5.3 Relationship between the ratio of potassium content in heartwood to sapwood (HW/SW) and the ratio of ^{137}Cs concentration in heartwood to sapwood (HW/SW). r, correlation coefficient; **, significant at 1% level

Cesium is a group I alkali metal that shows chemical properties similar to those of potassium; it is present in solution as a monovalent cation Cs^+ (White and Broadley 2000). Bruce and Richard (1993) and Iizuka et al. (2018) reported that ^{137}Cs movement from the sapwood to heartwood may be related to the heartwood:sapwood K content ratio. Altogether, these results confirmed that the radial movement of ^{137}Cs from sapwood to heartwood was affected by the K content gradient between them.

5.4 Conclusion

In this study, the radial distribution of ^{137}Cs concentration in stem wood of 33-year-old sugi trees was investigated 5 years after the Fukushima Daiichi Nuclear Power Plant accident. We arrived at the following main conclusions:

1. The mineral topsoil mean ^{137}Cs concentration in the surveyed area was 1727 Bq kg^{-1} (CV = 65%). Therefore, ^{137}Cs contamination of the topsoil in the surveyed area was quite heterogeneous.
2. ^{137}Cs in the bark and wood may be present in different chemical forms; in the bark, ^{137}Cs is insoluble and fixed, whereas in the wood it is soluble and mobile.
3. A highly significant and positive correlation was observed between the heartwood:sapwood ^{137}Cs concentration ratio and the corresponding K content ratio. A significant correlation was also found between the heartwood:sapwood ^{137}Cs concentration ratio and K content, moisture content, and L^* value in the heartwood.
4. Based on the results of the present study and other previously reported findings, we suggest that the movement of ^{137}Cs from sapwood to heartwood is very likely influenced by the K content gradient between them.

5.5 Materials and Methods

5.5.1 *Study Area*

Wood samples of Japanese cedar were collected from a plantation in the Funyu Experimental forest at Utsunomiya University (36°46′N, 139°49′E), in Shioya, Tochigi, Japan. The forest is located about 130 km southwest of the Fukushima Daiichi Nuclear Power Plant. According to the aircraft monitoring conducted by the Ministry of Education, Culture, Sports, Science and Technology, the degree of soil pollution at the Funyu Experimental Forest was estimated at 30–100 kBq m^{-2} on July 16, 2011 (MEXT 2011).

5.5.2 *Sample Collection, Processing, and Measurement*

The surveyed forest is composed of 33-year-old Japanese cedar trees planted on a near-flat area. A test area of about 0.15 ha (50 × 30 m) was set up in the forest where 80 trees corresponding to 30% of the total number of standing trees were cut down in February 2016. The mineral topsoil (0–5 cm) was sampled at 18 points throughout the test area using a 100-mL sampling cylinder (φ 50 × 51 mm, DIK-1801; Daiki, Saitama, Japan) to determine the ^{137}Cs concentration. A 5-cm-thick wood disk was sampled from the stem at a height of 0.2 m above the ground in all 80 trees.

The sampled surface mineral soils were sieved through a 2-mm mesh and sand gravels sized 2 mm or less were oven-dried and packed in a U-8 container (100 mL). The wood disks were cut into fan-shaped parts centered on the pith, and each part was further divided into heartwood, sapwood, and bark. The boundary between the heartwood and sapwood was confirmed visually. Each heartwood and sapwood sample was ground using a mill (IFM-S10G; Iwatani, Tokyo, Japan) to prepare wood dust for measuring ^{137}Cs concentration. The wood dust was oven-dried and packed in a U-8 container (100 mL). Oven-dried bark was minced into small pieces less than 5 mm in length with stainless scissors and packed in a U-8 container (100 mL). Sample ^{137}Cs (Bq kg^{-1} dry weight) concentration was measured with a germanium (Ge) semiconductor detector (Seiko EG & G, Ortec, Tokyo, Japan). Measurement conditions were as follows: measurement duration, 6000 s or longer, and gamma-ray peaks, 661.64 keV. Potassium (K) content (g kg^{-1} dry weight) in the heartwood and sapwood was measured by atomic absorption spectroscopy and determined with an atomic absorption photometer (Z-2310; Hitachi, Tokyo, Japan). Heartwood moisture content (%) was measured as $\{(Wg - Wo)/Wo\} \times 100$, where *Wg* is the weight of fresh wood and *Wo* is the weight of wood after drying in an oven at 105 °C until reaching constant weight. Heartwood color was measured by powdering the samples using a mill (IFM-S10G; Iwatani, Japan) and packing them in transparent plastic bags. The L^* value of heartwood under air-dry conditions was measured using a Chroma Meter (CR-400; Konica Minolta, Tokyo, Japan) and the $L^*a^*b^*$ (CIELAB) system.

Acknowledgment This research was financially supported by JSPS KAKENHI Grant numbers 24110001, 15K07496, and 17K00635. We would like to thank our many collaborators, Center for Bioscience Research and Education, and University Forest, Utsunomiya University, for assistance in this experiment.

Conflict of Interest The authors declare that they have no conflict of interest.

References

Abe Z, Oda K, Matsumura J (1994) The color change of sugi (*Cryptomeria japonica*) heartwood from reddish brown to black I, the color change and its causes. Mokuzai Gakkaishi 40:1119–1125. in Japanese with English summary

Bruce EC, Richard PG (1993) Anatomical, chemical, and ecological factors affecting tree species choice in dendrochemistry studies. J Environ Qual 22:611–619

Chigira M, Saito Y, Kimura K (1988) Distribution of Sr-90 and Cs-137 in annual tree rings of Japanese cedar, *Cryptomeria japonica* D Don. J Radiant Res 29:152–160

Iizuka K, Ohshima J (2017) Monitoring of ^{137}Cs concentration in stem wood of Japanese cedar in University Forests, Utsunomiya University. Suirikagaku 355:13–25. (in Japanese)

Iizuka K, Toya N, Ohshima J, Ishiguri F, Miyamoto N, Aizawa M, Ohkubo T, Takenaka C, Yokota S (2018) Relationship between ^{137}Cs concentration and potassium content in stem wood of Japanese cedar (*Cryptomeria japonica*). J Wood Sci 64:59–64

Imamura N, Konatus M, Ohashi S, Hashimoto S, Kajimoto T, Kaneko S, Takano T (2017) Temporal changes in the radiocesium distribution in forests over the five years after Fukushima Daiichi nuclear power plant accident. Sci Rep 7:8179

Ishiguri F, Maruyama J, Eizawa J, Saito Y, Iizuka K, Yokota S, Abe Z, Yoshizawa N (2006) Blackening phenomenon in heartwood of sugi originated from seedlings. Wood Ind 61:399–403. (in Japanese with English summary)

Iwase K, Tomioka R, Suriura Y, Kanasashi T, Takenaka C (2013) Absorption properties of the Cs in the bark of *Cryptomeria japonica* and *Quercus serrata*. Jpn J For Enviro 55:69–73. in Japanese with English summary

Kagawa A, Aoki T, Okada N, Katayama Y (2002) Tree-ring strontium-90 and cesium-137 as potential indicators of radioactive pollution. J Environ Qual 31:2001–2007

Katayama Y, Okada N, Ishimura Y, Nobuchi T, Aoki A (1986) Behavior of radioactive nuclides on the radial direction of the annual ring of sugi. Radioisotopes 35:636–638. in Japanese with English summary

Kawazumi K, Oda K, Tsutsumi J (1991) Heartwood properties of sugi (*Cryptomeria japonica*): moisture content of green wood, hot water extractives and lightness. Bull Kyushu Univ For 64:29–39. (in Japanese with English summary)

Kohno M, Koizumi Y, Okumura K, Mito I (1988) Distribution of environmental Cesium-137 in tree rings. J Environ Radioact 8:15–19

Kubo T, Ataka S (1998) Blackening of sugi (*Cryptomeria japonica* D. Don) heartwood in relation to metal content and moisture content. J Wood Sci 44:137–141

Kudo A, Suzuki T, Santry DC, Mahara Y, Miyahara S, Garrec JP (1993) Effectiveness of tree rings for recording Pu history at Nagasaki, Japan. J Environ Radioact 21:55–63

Kuroda K, Kagawa A, Tonosaki M (2013) Radiocesium concentrations in the bark, sapwood and heartwood of three tree species collected at Fukushima forests half a year after the Fukushima Dai-ichi nuclear accident. J Environ Radioact 122:37–42

Mahara Y, Ohta T, Ogawa H, Kumata A (2014) Atmospheric direct uptake and long-term fate of radiocesium in trees after Fukushima nuclear accident. Sci Rep 4:7121

Ministry of Education, Culture, Sports, Science and Technology (MEXT) (2011) Radiation dosage distribution map expansion site http://ramap.jmc.or.jp/map/. Accessed 07.07.2017 (in Japanese)

Miyamoto N, Iizuka K, Nasu J, Yamada H (2016) Genetic effects on heartwood color variation in *Cryptomeria japonica*. Silvae Genetica 65(2):80–87
Momoshima N, Bondietti EA (1994) The radial distribution of ^{90}Sr and ^{137}Cs in trees. J Environ Radioact 22:93–109
Morikawa T, Oda K, Matsumura J, Tsutsumi J (1996) Black-heartwood formation and ash content in the stem of sugi (*Cryptomeria japonica* D. Don) II, heartwood properties of three sugi cultivars. Bull Kyushu Univ For 74:41–49. (in Japanese with English summary)
Nagakura J, Abe H, Zhang C, Takano T, Takahashi M (2016) Cesium, rubidium and potassium content in the needles and Japanese cedar trees harvested from the sites of different radiocesium deposition levels. Jpn J For Environ 58:51–59. (in Japanese with English summary)
Nishikiori T, Watanabe M, Koshikawa MK, Takamatsu T, Ishii Y, Ito S, Takenaka A, Watanabe K, Hayashi S (2015) Uptake and translocation of radiocesium in cedar leaves following the Fukushima nuclear accident. Sci Total Enviro 502:611–616
Oda K, Matsumura J, Tsutsumi J, Abe Z (1994) Black-heartwood formation and ash contents in the stem of sugi (*Cryptomeria japonica* D. Don). Sci Bull Fac Agr Kyushu Univ 48:171–176. (in Japanese with English summary)
Ogawa H, Hirano Y, Igei S, Yokota K, Arai S, Ito H, Kumata A, Yoshida H (2016) Changes in the distribution of radiocesium in the wood of Japanese cedar trees from 2011 to 2013. J Environ Radioact 161:51–57
Ohashi S, Okada N, Tanaka A, Nakai W, Takano S (2014) Radial and vertical distributions of radiocesium in tree stems of *Pinus densiflora* and *Quercus serrata* 1.5 y after the Fukushima nuclear disaster. J Environ Radioact 134:54–60
Ohashi S, Kuroda K, Takano T, Sizuki Y, Fujiwara T, Abe H, Kagawa A, Sugiyama M, Kubojima Y, Zang C, Yamamoto K (2017) Temporal trends in ^{137}Cs concentration in the bark, sapwood, heartwood, and whole wood of four tree species in Japanese forests from 2011 to 2016. J Environ Radioact 178–179:335–342
Okada N, Katayama Y, Nobuchi T, Ishimaru Y, Yamashita H, Aoki A (1987) Trace elements in the stems of trees I. Radial distribution in sugi (*Cryptomeria japonica*). Mokuzai Gakkaishi 33:913–920
Okada N, Hirakawa Y, Katayama Y (2011) Application of activable tracers to investigate radial movement of minerals in the stem of Japanese cedar (*Cryptomeria japonica*). J Wood Sci 57:421–428
Okada N, Hirakawa Y, Katayama Y (2012) Radial movement of sapwood-injected rubidium into heartwood of Japanese cedar (*Cryptomeria japonica*) in the growing period. J Wood Sci 58:1–8
Tagami K, Uchida S, Ishii N, Kagiya S (2012) Translocation of radiocesium from stems and leaves of plants and the effect on radiocesium concentrations in newly emerged plant tissues. J Environ Radioact 111:65–69
Wang W, Hanai Y, Takenaka C, Tomioka R, Iizuka K, Ozawa H (2016) Cesium absorption through bark of Japanese cedar (*Cryptomeria japonica*). J For Res 21:251–258
White PJ, Broadley MR (2000) Mechanisms of caesium uptake by plants. New Phytol 147:241–256
Zhu YG, Smolders E (2000) Plant uptake of rediocaesium: a review of mechanisms, regulation and application. J Exp Bot 51:1635–1645

Chapter 6
Radiocesium Translocations in Bamboos

Mitsutoshi Umemura

Abstract In this chapter, the study (Umemura M, Kanasashi T, Sugiura Y, Takenaka C, J Jpn For Soc 97:44–50, 2015) on the distribution of radiocesium (^{137}Cs) in a Moso bamboo (*Phyllostachys pubescens*) forest in Fukushima Prefecture after the accident of Fukushima Daiichi Nuclear Power Plant (FDNPP) is introduced. We investigated the ^{137}Cs contamination levels in aboveground organs of the bamboos which sprouted before and after the accident and belowground organs, visually and quantitatively. From the analysis on the aboveground organs sampled in 2012, the highest ^{137}Cs concentrations were detected in the node parts of both the culms and branches of the bamboos sprouted before 2011 due to the direct fallout of radioactive substances. This fact indicates a long-lasting contamination without leaching of ^{137}Cs by rain which strongly adhered on the surface of the mature bamboos. From similar ^{137}Cs concentrations in each organ among the different-aged bamboos, it was supposed that ^{137}Cs diffused from the bamboos sprouted before 2011 to those sprouted in 2011 just after the accident and also that the root absorption of ^{137}Cs might affect the contamination in the bamboos sprouted in 2011. However, we did not find the evidence of root absorption in 2014 from the results of ^{137}Cs distribution in the root. These findings suggested that ^{137}Cs absorbed just after the accident has been diffusing throughout the bamboo forest via the rhizome system.

Keywords Bamboo · Radiocesium · Translocation · Surface absorption · Root absorption

6.1 Introduction

Moso bamboo (*Phyllostachys pubescens*) is an agriculturally and ecologically important plant species across East, South, and Southeast Asia. In parts of Japan, some landowners grow bamboo forests beside their houses to produce bamboo

---

M. Umemura (✉)
Hokkaido Research Center, Forestry and Forest Products Research Institute, Forest Research and Management Organization, Sapporo, Hokkaido, Japan

C. Takenaka et al. (eds.), *Radiocesium Dynamics in a Japanese Forest Ecosystem*,
https://doi.org/10.1007/978-981-13-8606-0_6

shoots. Since the severe accident of the Fukushima Daiichi Nuclear Power Plant (FDNPP) in March of 2011, the bamboo forests around the FDNPP have been contaminated by radioactive substances. Among the existing radioactive substances, radiocesium-137 (^{137}Cs) has a long half-life of 30.1 years and therefore can have long-lasting effects on the bamboo forests and thus human life. The contaminated forests might continue to discourage landowners to return and restart the production of bamboo shoots. To decontaminate the forests, proactive restoration activities such as removal of contaminated bamboos and reduction of air dose are needed.

Bamboo forests are unique in that each of them consists of a limited number of bamboo individuals that are connected with belowground rhizomes (i.e., genets). The rhizomes transport various types of materials such as nutritional elements and assimilation products from mature culms to new shoots. Understanding how ^{137}Cs is transported via these rhizomes provides a critical step toward removing the radioactive substances from both the above- and belowground parts of bamboos.

We still have limited knowledge about the pathways through which ^{137}Cs move within bamboo individuals and in bamboo forest ecosystems. A previous study has found within-individual gradients of radiocesium concentrations (^{134}Cs and ^{137}Cs) in Moso bamboo shoots, where the upper sections of the edible part had higher concentrations (Higaki et al. 2014). Such within-individual localizations of radioactive substances have also been visually observed using autoradiograph approaches (Minowa 2013). It is yet unclear whether radiocesium in bamboo shoots is directly absorbed from the soil via roots or translocated from other culms via rhizomes. Moso bamboos often form “root mats” close to the soil surface (ca. 0–10 cm depth), and it is possible that the radiocesium has been actively absorbed through these roots after the accident. On the other hand, it is known that many tree species can absorb radiocesium into their body not only from their roots but also from their leaves and barks (Sato 2012). Thus, it is also possible that the radiocesium adhered on the surface of mature bamboo culms would be absorbed into their inner tissues and further transported to new shoots via rhizomes, taking into account the fact that the growths of shoots are often supported by phloem transportations of assimilation products from mature culms. These possibilities, however, have not been tested.

In this section, we introduce our previous study (Umemura et al. 2015) about root absorption, culm-surface absorption, and translocation of ^{137}Cs after the accident. We also discuss possible ways to efficiently decontaminate the bamboo forests so that the land owners can produce bamboo shoots again.

6.2 The Surface Adhesion of Radioactive Substances on the Aboveground Organs

In May and June 2012, we investigated the distribution of ^{137}Cs in the Moso bamboo forest located near a household in Kawamata Town, Date district, northeastern Japan (Fig. 6.1). Specifically, we analyzed ^{137}Cs concentrations in culms, branches, and

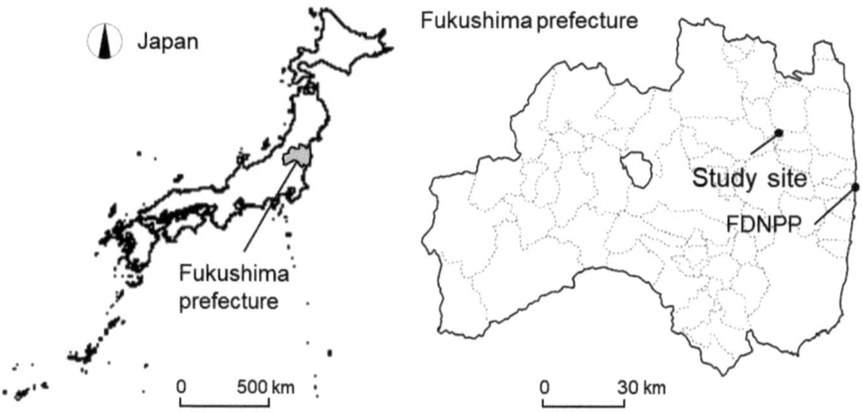

Fig. 6.1 Location of the study site in Fukushima prefecture, northeastern Japan. The Moso bamboo forest at this site is located about 35 km northwest of the Fukushima Daiichi Nuclear Power Plant (FDNPP) and on a southeast-facing slope facing toward the FDNPP. The air dose of the forest in May 2012 (ca. 14 months after the accident) was 4.5 $\mu Sv\ h^{-1}$ (1 m above the ground)

leaves of the bamboos which sprouted before and during 2011, a bamboo shoot which sprouted in 2012, and a rhizome (Fig. 6.2a). It is important here to note that the bamboos sprouted before 2011 had existed from before the accident and thus were exposed to the fallout of radioactive substances. Bamboos sprouted in 2011, on the other hand, were not exposed because bamboos generally sprout in April to May; i.e., the bamboos in 2011 sprouted after the accident. At first, we visually observed the distribution of radioactive substances that should have adhered on the culms, branches, and leaves of the bamboos sprouted before 2011 by an autoradiography method with imaging plate (IP). To investigate how much radioactive substances firmly adhered, we observed those samples by two kinds of washing methods: (1) only distilled water rinsing and (2) brushing and ultrasonic washing + distilled water rinsing.

As the results, a lot of black spots that are showing high radioactivity were found on/in the culms and branches, in particular, node sections of them (Fig. 6.3a). Majority of those radioactive substances on the nodes were removed by physical washing using a brush, but those on the nodes of branches were hardly removed by the same way. For the leaves, radioactive substances were observed along the shape of the leaves, indicating that the radioactive substances were taken into the leaves.

High concentrations of ^{137}Cs were found at the node sections of culm surface of the bamboos sprouted before 2011 (Fig. 6.3b). The concentrations were especially higher in the upper nodes than in the nodes of the culm base. Furthermore, the ^{137}Cs concentrations declined by brush-washing. These results show that ^{137}Cs removed was clearly derived from substances adhering on the culm surface and which should be brought by the direct fallout. On the other hand, from the ^{137}Cs analysis of the bamboos sprouted in 2011, there was almost no difference of ^{137}Cs concentrations between two kinds of washing methods. This indicates that there was hardly adhesion of ^{137}Cs on the culm surface (Fig. 6.3b).

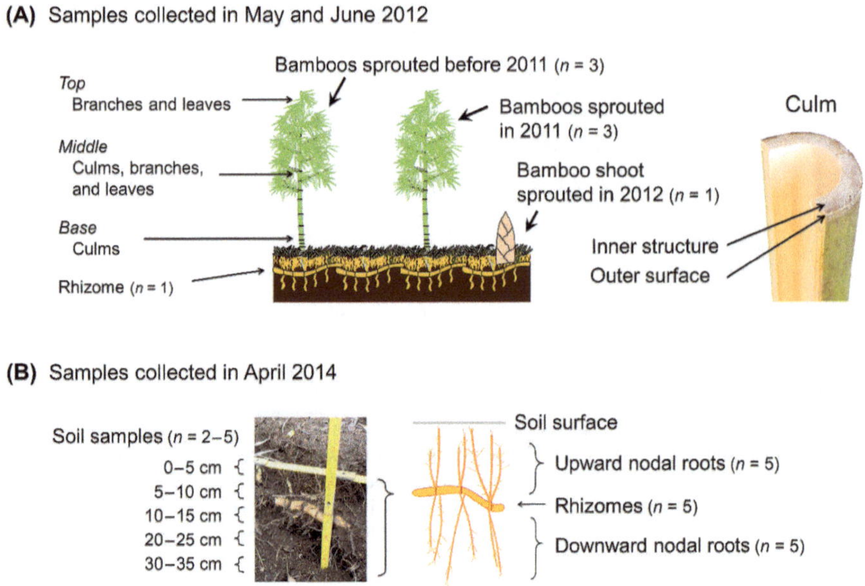

Fig. 6.2 Above- and belowground parts of Moso bamboos and soil sampled in this study. The culm walls were separated into two sections: (1) green-colored tissues in the outermost surface (ca. 1–2 mm thick) and (2) the inner structure of culms, where the vascular bundles are distributed, between the inner and outer surfaces (ca. 10–15 mm thick) (note that bamboo culms are hollow; the inner surface refers to the tissue facing the hollows). (This figure was modified based on Umemura et al. 2015)

From our observation of the culms, branches, and leaves of the bamboos sprouted before 2011, it was revealed that the radioactive substances were distributed especially at high concentrations on the nodes of the culms and branches. This seems due to the adhesion of radioactive substances to "waxy materials" around the nodes. In general, the waxy materials are gradually firmly fixed to the culm surface along with the age, changing from white to black (Fig. 6.3c). The waxy materials are also in a fibrous or powdery form. In previous study, a part of radioactive substances that were brought by fallout immediately after the FDNPP accident is considered to have dissolved in rainwater or fallen as particulates (Kaneyasu et al. 2012; Yamaguchi et al. 2012; Adachi et al. 2013). Therefore, the waxy materials may have had caught water-soluble radioactive substances dissolved in stem flow, and/or the fine particles supplied from atmosphere.

It is important here to note that radioactive substances adhering on nodes of the branches were hardly removed by the physical washing using the brush, although those on the culm scarcely removed (Fig. 6.3a). These results also suggest that radioactive substances which adhered especially on the nodes are rarely washed off by rain and remaining on the culm surface for a long time.

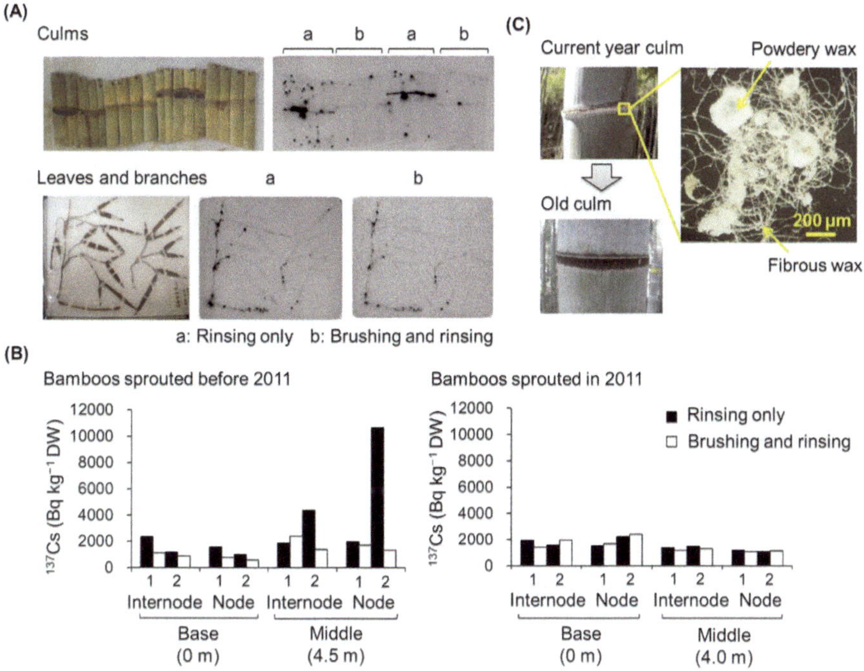

Fig. 6.3 (**a**) Radiographs of Moso bamboos that sprouted before 2011 (before the FDNPP accident). Radioactive substances were washed off using two methods: rinsing only and brushing and rinsing combined (the time exposed on imaging plate (IP) = 39–56 h). (**b**) ^{137}Cs concentrations in outer surfaces of the culms that sprouted before 2011 and in 2011 and washed by two different methods. The numbers on the *x*-axis (1 and 2) indicate sample numbers. (**c**) Microscope image of waxy materials on the nodes of bamboo culms. (This figure was modified based on Umemura et al. 2015)

6.3 A Possibility of the Surface Absorption of ^{137}Cs

The radioactive substances, which adhered on the bamboo surface, are likely to have absorbed from the surface into the body and subsequently translocate to the other organs after the FDNPP accident, given the fact that ^{137}Cs should contain water-soluble forms to be easily taken into plants then. Here, we hypothesized that ^{137}Cs concentrations in the bamboos sprouted before 2011 (i.e., before the accident) were higher than those in the bamboos sprouted in 2011 (i.e., after the accident). As the results, against the hypothesis, the bamboos that had sprouted before 2011 and during 2011 had similar concentrations of ^{137}Cs in culms, branches, and leaves (Fig. 6.4). Since the bamboos sprouted in 2011 were not exposed to the fallout, ^{137}Cs which was contained in those aboveground organs should have been transported from belowground organs by the root absorption and/or translocation via rhizomes, not from the surface absorption on the bamboos sprouted in 2011.

To reveal sources of ^{137}Cs transported from belowground organs, we will discuss their possible pathways based on the vertical and horizontal distribution

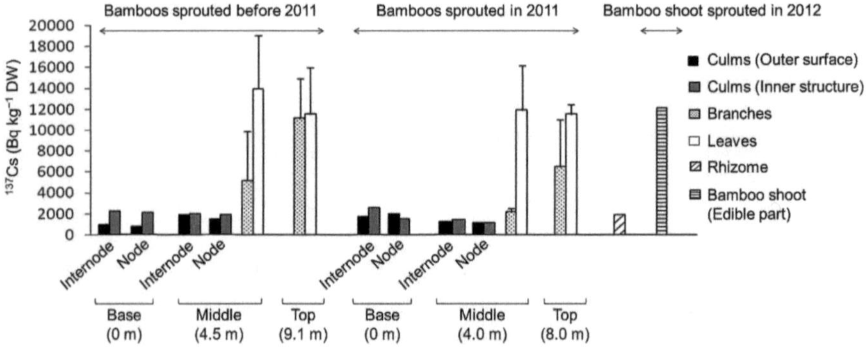

Fig. 6.4 ^{137}Cs concentrations in different organs of Moso bamboos sampled at different heights. Error bars indicate standard deviations (culms, $n = 2$; branches and leaves, $n = 3$; rhizome and bamboo shoot, $n = 1$). (This figure was modified based on Umemura et al. 2015)

patterns of ^{137}Cs in the culms, where we found slightly different patterns of ^{137}Cs between before 2011 and during 2011 (Fig. 6.4). In the bamboos sprouted in 2011, ^{137}Cs concentrations in the culms tended to be higher at the base parts in comparison with at middle height. This tendency was similar to potassium (K) distribution which was higher at culm base (3.7–14 mg g^{-1}) than at 5 m height (1.2–3.2 mg g^{-1}) based on a previous study which was conducted in another bamboo forest (Umemura 2014). This indicates that ^{137}Cs was supplied from belowground organs such as rhizome or roots. On the other hand, such tendency was not found in the bamboos sprouted before 2011, where ^{137}Cs concentrations were higher in inner structure than in outer surface (t-test $n = 8$, $p < 0.05$). The tendency was especially remarkable at the base parts. These characteristic distributions of ^{137}Cs might result from the surface absorption on the culm of the bamboos sprouted before 2011.

Regarding the surface absorption of radioactive Cs in plants, some previous studies reported that the leaf absorption affects the dynamics of radioactive substances in trees, especially immediately after nuclear power plant accident (Calmon et al. 2009), and that in peach trees, radioactive Cs that was absorbed from the bark was detected in the fruits (Takata 2013). Though there is no detailed verification of the mechanism of surface absorption of ^{137}Cs in bamboo plant, it is important to clarify the chemical compounds which were absorbed into inner structure, and their translocation pathway to the inner tissue from the culm surface, understanding their structural characteristics.

6.4 Root Absorption of ^{137}Cs After the FDNPP Accident

In this study, we suggested that ^{137}Cs which was detected at high concentrations in the bamboos sprouted in 2011 should have been transported from belowground organs by the root absorption and/or translocation from mature culms. To discuss which is the source of ^{137}Cs accumulated, root absorption or translocation from mature culms, we made following two hypotheses from a point of view of periods during which ^{137}Cs could be supplied to the bamboos sprouted in 2011. One of the possible periods is right after the accident when ^{137}Cs should have been supplied by root absorption or via rhizomes from the mature bamboos sprouted before 2011. Another one is a period throughout a year after the accident during which ^{137}Cs could be continuously supplied to the bamboos sprouted in 2011.

To discuss these hypotheses, we would estimate the ^{137}Cs dynamics based on the potassium (K) ones in Moso bamboo, assuming that the behaviors of Cs and K are similar in the bamboo plant. In a previous study, K concentrations in whole culms of Moso bamboo continued to decrease by a dilution effect with the rapid growth until 6 weeks after the bamboo sprouting (Wu et al. 2009). The researchers also found that K concentrations in the culms and branches continued to decrease with age. Therefore, it is supposed that Moso bamboo does not continue to accumulate ^{137}Cs throughout the year after the bamboo sprouting. Furthermore, in another previous study, K concentrations in the leaves decreased from September to the next March, indicating retranslocation of K from leaves to culms or rhizomes (Umemura and Takenaka 2014b). Additionally, K concentrations in the first- and second-aged rhizomes of Madake (*Phyllostachys bambusoides*) seasonally increased during the period from July to October and in March (Ueda et al. 1961). From these two reports, it is indicated that most of K accumulated in the bamboo shoots is considered to derive from one which has been stocked in rhizomes by re-translocation from mature culms. Given these characteristics of K dynamics in the bamboo plant, ^{137}Cs contained in Moso bamboos sprouted in 2011 seems not to have been accumulated throughout the year after the sprouting but to have been supplied in the bamboo shoots via rhizomes during the shoot sprouting and/or by root absorption right after the accident.

In the middle of March 2011 (right after the accident) when a large amount of ^{137}Cs was deposited on mature culm surface, it is supposed that there may be not so much translocation of ^{137}Cs to the rhizomes from the contaminated mature culms because at that time retranslocation of K from leaves seems to be calm down considering the seasonal change in leaves as mentioned above (Umemura and Takenaka 2014b). Therefore, it is considered that all of the ^{137}Cs accumulated in the bamboo shoots did not necessarily derive from radioactive substances adhered on the surface of mature culm; root absorption of ^{137}Cs should have also occurred. Right after the accident, ^{137}Cs contained some water-soluble forms such as sulfate aerosol (Kaneyasu et al. 2012). In another report in *Houttuynia cordata*, a kind of the perennial herbs, high concentrations of ^{137}Cs in 2011 were detected in the current leaves that did not yet foliate at the time of the accident (Sugiura et al. 2016). This observation clearly shows that the herbs absorbed ^{137}Cs via roots of the belowground part in the spring of 2011. Therefore, also in bamboo forests, it is enough to consider that Moso bamboo had taken up readily available ^{137}Cs which should have deposited on the soils right after the accident.

6.5 Distribution of ^{137}Cs Belowground and the Root Uptake in 3 Years After the Accident

In order to verify whether ^{137}Cs was absorbed by the root in 3 years after the accident, we evaluated a hypothesis that if ^{137}Cs was absorbed by the root, ^{137}Cs concentrations can be higher in roots elongating upward from node of rhizomes (upward nodal roots) than in roots elongating downward from there (downward nodal roots) because the upward nodal roots should be exposed to a high level contamination of ^{137}Cs at the surface of soil. To test this hypothesis, we sampled soils from 0 to 35 cm depth, rhizomes (including a few nodes), and those vertical nodal roots (especially primary roots) in 2014 (Fig. 6.2b). Soils adhered on the rhizomes and the roots were carefully removed by ultrasonic and brush washing for ^{137}Cs analysis. As the results, ^{137}Cs concentrations in the soil showed the highest value (19,000 Bq kg^{-1} DW–93,000 Bq kg^{-1} DW) at the surface layer 0–5 cm even in 3 years after the accident (Fig. 6.5a). ^{137}Cs concentrations in the rhizomes, which were distributed within 3–13 cm of soil depth, were 196–1216 Bq kg^{-1} DW, and the variation did not relate with ^{137}Cs concentrations in soil along the depth (Fig. 6.5b). Moreover, there was no significant correlation ($P > 0.05$) between the ^{137}Cs concentrations of individual roots and rhizomes ($R = 0.50$) and no tendency of ^{137}Cs between upward roots and downward roots (Fig. 6.6). These results indicate that uptake of ^{137}Cs from roots is very small in 2014.

In general, Moso bamboo forest develops root mat layer on the surface soil, where 54 to 60% of the whole root biomass is existing within 0–10 cm of soil depth (Umemura and Takenaka 2014a). Therefore, the bamboo can actively absorb various nutrients from there. In our study, however, we could not clearly confirm the root absorption of ^{137}Cs. This is perhaps because ^{137}Cs was present in stable fractions not in available ones in the soil. According to a previous study regarding fallout ^{137}Cs derived from the Chernobyl accident, Tsukada et al. (2008) reported that in decades

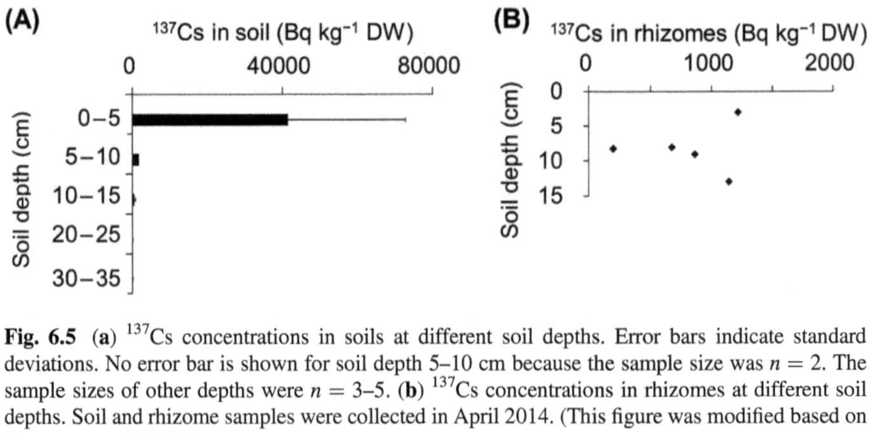

Fig. 6.5 (**a**) ^{137}Cs concentrations in soils at different soil depths. Error bars indicate standard deviations. No error bar is shown for soil depth 5–10 cm because the sample size was $n = 2$. The sample sizes of other depths were $n = 3$–5. (**b**) ^{137}Cs concentrations in rhizomes at different soil depths. Soil and rhizome samples were collected in April 2014. (This figure was modified based on Umemura et al. 2015)

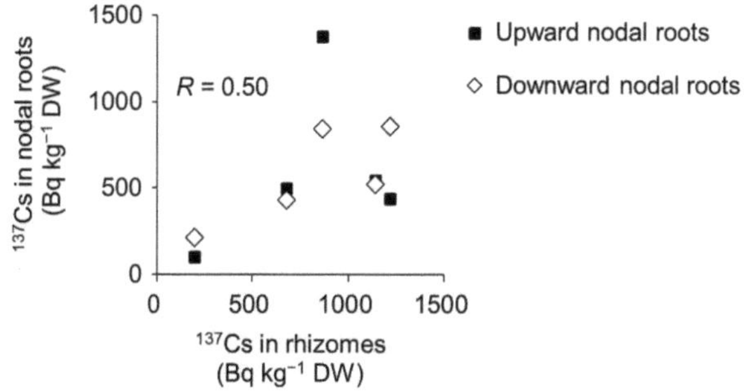

Fig. 6.6 Relationship of ^{137}Cs concentrations between rhizomes and upward or downward nodal roots (sampled in April 2014). *R* indicates the Pearson's product moment correlation coefficient (calculated using all the upward and downward nodal root samples; $n = 10$). (This figure was modified based on Umemura et al. 2015)

after the accident, ^{137}Cs in soils was contained at the rate of 10% in the exchangeable, 20% in organic-bound fractions, and 70% in the strongly bound fraction such as clay mineral. In addition, Tsukada (2014) reported that ion-exchange fractions of ^{137}Cs in soils, which was added to soils (Andosol) and extracted with 1 M ammonium acetate, decreased to ca. 60%, 34%, 25%, and 22% after 2 days, 64 days, 1 year, and 2 years, respectively. Therefore, at the time of 2014, when 3 years have passed since the FDNPP accident, it is considered that most ^{137}Cs existed in a stable fraction in the soil and that their root absorption from the soil was limited.

We also found that the ^{137}Cs concentrations in rhizomes did not depend on ^{137}Cs contamination level in every depth of the soil (Fig. 6.5b). In addition, the concentrations of ^{137}Cs in rhizomes and roots were almost similar (Fig. 6.6). These results indicate that ^{137}Cs which had been taken into the body immediately after the accident has diffused in the rhizomes and roots at the time of 2014. It is important here to note that ^{137}Cs concentrations in those rhizomes or roots were relatively low among the various types of bamboo organs sampled in this study. This is likely to attribute to the sampling season of their organs. Various nutrients in the rhizomes could be most consumed for the new bamboo shoot growth in the end of April when we sampled rhizomes and roots. This may be why low ^{137}Cs concentrations found in the rhizomes and roots. And also, ^{137}Cs concentrations in the culms and rhizome sampled in May and June in 2012 may reflect those that remained there after the translocation. In order to clarify the dynamics of Cs in the aboveground and belowground parts of Moso bamboo, further investigation is needed with considering the seasonal variation.

6.6 Conclusions

We studied the distribution of ^{137}Cs in a Moso bamboo forest after the FDNPP accident (Umemura et al. 2015). Specifically, we investigated ^{137}Cs concentration in different organs of bamboos and soil in 2012 and 2014 (i.e., 1 and 3 years after the accident). In 2012, highest radioactivity was found in the nodes of culms and branches of the bamboos that had sprouted before the accident. We also found that the radioactive substances adhered strongly to the bamboo-originating waxy materials around the nodes, thereby staying free from being washed off by rain. These results suggest that the culms that had been exposed to direct fallouts can lead to long-term contaminations of the forest. In addition, we found that the bamboos that had sprouted before 2011 and during 2011 (i.e., before and after the accident) had similar concentrations of ^{137}Cs in culms, branches, and leaves. This result indicates that ^{137}Cs was transported into the newly sprouted bamboos after the accident through root absorptions and/or rhizome-mediated translocations from the bamboos that had sprouted before 2011. Assuming that ^{137}Cs is translocated in a similar manner with potassium, it is possible that the ^{137}Cs in the new bamboos was translocated from the old bamboos, the surface of which ^{137}Cs was absorbed from. In addition, ^{137}Cs in the new bamboos could have been absorbed from the roots, given the fact that the ^{137}Cs having readily available chemical forms were emitted from the FDNPP just after the accident. In 2014, however, we found no evidence for root absorptions. These results suggest that ^{137}Cs was absorbed immediately after the accident and diffused across the bamboo forest via rhizome networks, while later on, hardly no ^{137}Cs was further absorbed from the roots.

Our study provides important implications to decontaminating the bamboo forests. We found that the per capita concentration of the ^{137}Cs that has been already absorbed in the bamboo plants would likely gradually decrease due to translocations to new shoots. Nevertheless, this does not necessarily mean that the absolute amount of ^{137}Cs in the ecosystem will decrease faster than its half-life time, given that the ^{137}Cs in culms and leaves can be reabsorbed through roots after falling and being decomposed. It is also possible for the rhizomes and roots to spread into the nearby non-bamboo forests and transport additional ^{137}Cs into the bamboo forests. Our results suggest that, in order to enhance the decontamination and reduction of air dose in the bamboo forests, we should proactively exclude the newly sprouting bamboo shoots in which ^{137}Cs can accumulate at high concentrations. It would also be effective to thin the old bamboos on which radioactive substances would adhere and to remove the litters from the forest floor. We believe that such restoration activities would help decontaminate the bamboo forests and encourage the return of landowners.

References

Adachi K, Kajino M, Zaizen Y, Igarashi Y (2013) Emission of spherical cesium-bearing particles from an early stage of the Fukushima nuclear accident. Sci Rep 3:2554

Calmon P, Thiry Y, Zibold G, Rantavaara A, Fesenko S (2009) Transfer parameter values in temperate forest ecosystems: a review. J Environ Radioact 100:757–766

Higaki T, Higaki S, Hirota M, Hasezawa S (2014) Radiocesium distribution in bamboo shoots after the Fukushima nuclear accident. PLoS One 9:e97659

Kaneyasu N, Ohashi H, Suzuki F, Okuda T, Ikemori F (2012) Sulfate aerosol as a potential transport medium of radiocesium from the Fukushima nuclear accident. Environ Sci Technol 46:5720–5726

Minowa H (2013) View radiation from pine, bamboo, and plum – analysis using imaging plate. Housha Kagaku 27:45–52. (in Japanese)

Sato M (2012) Fruit in Fukushima in the nuclear accident year – report of the first year examination to take measures against the radioactive contamination of nuclear power plant accident. Radiochem News 26:21–31. (in Japanese)

Sugiura Y, Shibata M, Ogata Y, Ozawa H, Kanasashi T, Takenaka C (2016) Evaluation of radiocesium concentrations in new leaves of wild plants two years after the Fukushima Dai-ichi nuclear power plant accident. J Environ Radioact 160:8–24

Takata D (2013) Distribution of radiocesium from the radioactive fallout in fruit trees. In: Nakanishi TM, Tanoi K (eds) Agricultural implications of the Fukushima nuclear accident. Springer, Japan, pp 143–162

Tsukada H (2014) Behavior of radioactive cesium in soil with aging. Jpn J Soil Sci Plant Nutr 85:77–79. (in Japanese)

Tsukada H, Takeda A, Hisamatsu S, Inaba J (2008) Concentration and specific activity of fallout ^{137}Cs in extracted and particle-size fractions of cultivated soils. J Environ Radioact 99:875–881

Ueda K, Ueda S, Yakushiji K (1961) On the seasonal changes of the nutrient contents in Madake (*Phyllostachys reticulata*). Bull Kyoto Univ For 33:55–66. (in Japanese)

Umemura M (2014) Biogeochemical dynamics of nutritional elements in Moso bamboo (*Phyllostachys pubescens*) forests (doctoral dissertation). Nagoya University, Nagoya, p 111

Umemura M, Takenaka C (2014a) Biological cycle of silicon in Moso bamboo (*Phyllostachys pubescens*) forests in central Japan. Ecol Res 29:501–510

Umemura M, Takenaka C (2014b) Retranslocation and localization of nutrient elements in various organs of Moso bamboo (*Phyllostachys pubescens*). Sci Total Environ 493:845–853

Umemura M, Kanasashi T, Sugiura Y, Takenaka C (2015) The distribution of radiocesium in a Moso bamboo (*Phyllostachys pubescens*) forest in Fukushima Prefecture. J Jpn For Soc 97:44–50. (in Japanese)

Wu J, Xu Q, Jiang P, Cao Z (2009) Dynamics and distribution of nutrition elements in bamboos. J Plant Nutr 32:489–501

Yamaguchi N, Takada Y, Hayashi K, Ishikawa S, Kuramata M, Eguchi S, Yoshikawa S, Sakaguchi A, Asada K, Wagai R, Makino T, Akahane I, Hiradate S (2012) Behavior of radiocesium in soil-plant systems and its controlling factor. Bull Natl Inst Agro-Environ Sci 31:75–129. (in Japanese)

Chapter 7
Movement of Cesium in Model Plants

Jun Furukawa

Abstract The identification of mechanisms regulating ^{137}Cs transport from the rhizosphere to plants and between plant tissues such as root and shoot is highly necessary. For investigating such ^{137}Cs transport in the crops and trees, the experiments using model plants are carried out under the controlled conditions. The advantages such experiments are application of genome information and comparison of the effects induced by the different environmental conditions. In this chapter, the recent knowledge obtained from some crops and trees are introduced, and the insight for the further ^{137}Cs transport research is explained.

Keywords ^{137}Cs transport · Cultivation condition · Model plants · Substrate specificity

7.1 Absorption and Translocation of Radioactive Cesium in Model Crops

To clarify the absorption and translocation processes of radioactive cesium (Cs) in crops, the experiments performed under controlled conditions like an experimental field or laboratory are also important. The acquisition of insight about the mechanisms involved in the absorption and translocation of radioactive Cs from the soil into the plant using model crops will provide the evidences for the local measures and contribute to the suppression of radioactive Cs absorption into the plant.

The results of field experiment conducted at paddy fields in Fukushima City in 2011 and 2012 revealed that the content of radioactive Cs in rice straw and brown rice has a large difference between rice varieties. This result shows that the traits related to Cs uptake and translocation to brown rice were differed between rice varieties and it is possible to produce low Cs absorption rice by elucidating the responsible gene(s) by molecular biological techniques (Ohmori et al. 2014a). In the

J. Furukawa (✉)
Faculty of Life and Environmental Sciences, University of Tsukuba, Tsukuba, Japan
e-mail: furukawa.jun.fn@u.tsukuba.ac.jp

C. Takenaka et al. (eds.), *Radiocesium Dynamics in a Japanese Forest Ecosystem*,
https://doi.org/10.1007/978-981-13-8606-0_7

similar experiment, investigation of the fertilization effect showed that the increase of the nitrogen fertilizer amount under low potassium (K) condition increased the amount of radioactive Cs contained in rice straw and brown rice up to almost twice (Ohmori et al. 2014b). Generally, K, which is considered to compete with Cs during the uptake and translocation into plants, was investigated focusing on the amount in the soil in many reports; however, the finding that nitrogen application affects Cs accumulation in brown rice provides the new aspect, nitrogen nutrition, into the Cs behavior research in plants.

In 2017, Ishikawa et al. reported a new Cs uptake regulation mechanism related to sodium (Na) efflux from cytosol (Ishikawa et al. 2017). Plants grown under low K condition accumulate Na in their bodies instead of K. Plants containing excess amount of Na activate salt overly sensitive (SOS) control pathway to protect from salt stress. In the SOS regulatory pathway, three SOS genes (*SOS1*, *SOS2*, and *SOS3*) play a central role. The activity of SOS1, Na^+/H^+ antiporter, is regulated by SOS2-SOS3 complex, and excessive intracellular Na^+ is excluded to the outside of the cell. Ishikawa found that the *OsSOS2* is a responsible gene of low Cs absorption rice produced by ion beam-induced mutagenesis. Because of the functional deficiency of OsSOS2, intracellular Na^+ is not excluded; therefore, intracellular Na concentration increases. Under such a situation, it is predicted that intracellular high Na concentration or Na^+ itself inhibits the expression of Cs (originally K) uptake-related genes (such as *HAK* and *AKT*) suggesting that the involvement of mechanisms regulates Na and K homeostasis within the cytosol.

For producing low Cs absorption rice, analysis of a series of rice mutants with reduced Cs absorption was also carried out. All the mutants which showed low Cs uptake had the same responsible gene, *OsHAK1* (Rai et al. 2017). This low Cs absorbed rice was succeeded in reducing the Cs amount to 1/10 or less comparing to the parent cultivar (Fig. 7.1). Moreover, since this low Cs absorption rice showed no change in the K content, it suggests that the main Cs uptake process in rice cultivated under the low K concentration environment is carried by OsHAK1. In addition to the conventional mutagenesis approach, low Cs absorption rice was produced by genome editing technology using CRIPR-Cas system which was remarkably developed recently (Nieves-Cordones et al. 2017). The focused gene was also *OsHAK1*, and the plants losing the function of OsHAK1 showed that the Cs absorption capacity was markedly reduced. Focusing on Cs transport activity via HAK family transporters, at least two independent functional regions of the HAK transporter protein explain the Cs transport activity (Alemán et al. 2014; Rai et al. 2017). Based on those researches, it is expected that investigation of the amino acid sequences of these sites is effective for the production and selection of low Cs absorbed plants.

In the monitoring inspection for radioactive substances in Fukushima prefecture, radioactive Cs concentration in almost all agricultural and livestock products was kept lower than the standard of radioactive Cs level (100 Bq/kg) after 2012, and most of crops, including rice, exceeded the standard level which was hardly observed. However, in soybean, there were some cases where the radioactive Cs was accumulated over the standard level after 2013 (Nihei 2016). Since the seeds of legume crops often show the high K content, an investigation of Cs accumulation in *Lotus*

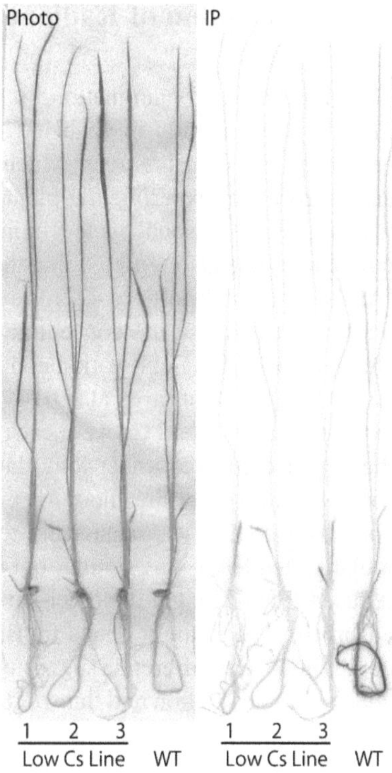

Fig. 7.1 Image of radioactive cesium (^{137}Cs) uptake by rice seedlings. Left panel shows the photo of each rice and right shows the ^{137}Cs distribution obtained by Imaging Plate (IP). In ^{137}Cs distribution image, color change from white to black indicates ^{137}Cs accumulation

japonicus was carried out at the laboratory using various levels of K concentration. *L. japonicus* is a model legume, which indicates its genome was already sequenced, the molecular biological approach is applicable, and the bio-resources are easily available. Cesium uptake analysis using hydroponically cultivated *L. japonicus* indicated Cs absorbed was mainly accumulated in root, and translocation ratio of the root absorbed Cs to the shoot part was not differed among cultivars. However, the concentration of root Cs was different between the cultivars. It suggests that it is possible to reduce Cs absorption in legumes, including soybean, if the responsible gene(s) is identified like a rice case. In addition, K deficiency-induced Cs uptake was also varied depending on the cultivars. *L. japonicus* grown under low K condition was expected to enhance K absorption activity, and the absorption of Cs would also be promoted. However, K deficiency-induced increase of Cs uptake was differently regulated between the cultivars and the one cultivar enhanced Cs uptake and the other was not. In the cultivar in which the influence of the K deficiency hardly appeared, the gene expression analysis also revealed that the expression of the transcription factor reported to be induced by K deficiency in *Arabidopsis thaliana* was not increased. Therefore, in *L. japonicus*, it was strongly suggested that the mechanism respond to K deficiency was differed from *Arabidopsis* and differed also between cultivars.

7.2 Absorption of Radioactive Cesium in a Model Tree

The influence of contamination by radioactive materials is large in the forest area. Therefore, elucidation of Cs behavior in the forest ecosystem and woody plants is necessary. For focusing trees, which grow over several years or decades, poplar was employed as an experimental plant because of the established cultivation method with shortened seasonal cycle within the laboratory. Poplar is not only a widely used model tree; it is also reported that the Cs transfer activity to the leaves is high in the IAEA technical report No.472. Previous studies investigating elemental profiles in poplar revealed that the concentration of phosphorus and K in mature leaves was changed by shifting cultivation conditions from long-day length condition to short-day condition (Kurita et al. 2014; Furukawa et al. 2012). Therefore, the uptake experiments of K and Cs were carried out under long- and short-day conditions. In the poplar cultivated under long-day conditions, radioactive Cs absorbed from the roots was moved to the shoot apex and the concentration of Cs was highest at the apex. However, after transferring to the short-day condition, the shoot apex accumulating tendency was similar, but the total amount of absorbed Cs was drastically reduced (Fig. 7.2). On the other hand, under short-day conditions, the amount of K absorbed through the root and transported to the shoot did not differ from long-day conditions. That difference suggested that dormancy induced by short-day treatment changed an absorption and transport processes of Cs and the pathway involved in the Cs transport was independent from K transport system (Noda et al. 2016).

7.3 Radioactive Cesium Recovery from Mature Leaves Before Fallen Leaves in Trees

When dormancy is induced in trees with short-day length, K concentration in mature leaves decreases (Furukawa et al. 2012). This phenomenon is thought to be due to the nutrient recovery mechanism before fallen leaves, suggesting the possibility that Cs accumulated in the leaves will be also translocated to the trunk by the K pathway and be kept within the tree body for a long period. In order to identify this recovery process, the Cs behavior in the tree was analyzed under long- and short-day conditions by applying Cs to a mature leaf. In long-day condition poplar, most of Cs was transported to the tissue below the Cs-applied leaf, and in the short-day condition, Cs was transported to the tissue above the treated leaf. In general, substrate transport through the xylem occurs from root to leaf according to the transpiration flow; however, when the substrate is transported via sieve tubes, it is transferred to the required tissue. Therefore, the change in Cs transport direction observed under short-day condition was regulated by the change of K required tissue through the sieve transport and/or the exchange of major pathway of Cs (originally K) from sieve to xylem transport around the node. Based on the pathway exchange idea, the expressions of several K transporters were compared between long- and

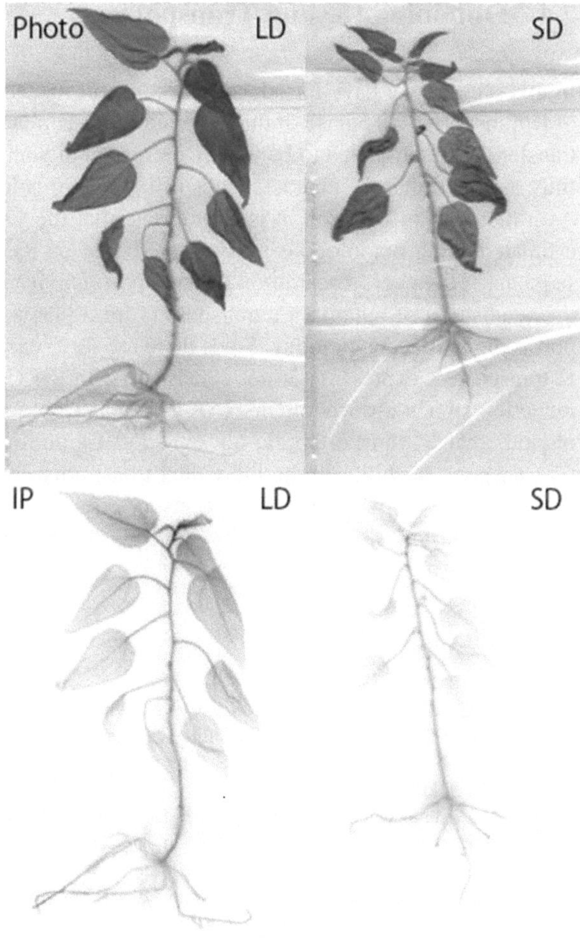

Fig. 7.2 Localization of root-applied ^{137}Cs under long-day (LD) and short-day (SD) length conditions. The upper panel shows the photo and the lower shows the ^{137}Cs distribution obtained by Imaging Plate (IP). Bars in IP image show 1 cm and color change from white to black indicates ^{137}Cs accumulation

short-day conditions, and among these K transporters, one of the efflux type transporters was highly upregulated around the node. For identifying its involvement for the change of Cs transport direction, the recombinant poplar suppressed its expression was produced, and Cs transport under short-day condition was analyzed. As a result, the Cs behavior was not changed under short-day condition in suppressed poplar, and the involvement of that transporter into the Cs translocation was confirmed. Changes in the Cs dynamics in the short-day transition are thought to be induced by the change of K required tissue in preparation for overwintering or spring budding, and radioactive Cs might be also circulated within the tree body by the seasonal expression patterns of K transporters. For the prediction of Cs behavior within trees and forest ecosystems, it is necessary for understanding the similar Cs transport mechanisms among different tree species and between seasons.

7.4 Obtaining Cesium Transporters

In this section, we have introduced Cs transport mechanisms through several types of K transporters. Because of the large family of K transporters, the transporters that transfer Cs are predicted to be widely preserved in plant species. However, it is also fully assumed that the characteristics of homologous transporters vary one by one. As a first step of research, it is appropriate that the Cs transport activity has been estimated using homologous transporters based on its similarity of the amino acid sequence. However, research on Cs transport activity in each K transporter and/or the acquisition of putative Cs transporters from plants with characteristic Cs transport is highly necessary today. Since there are some cases that the behavior of K and K transporters expression is not consistent with the Cs transport, as shown in the reduction of Cs absorption in HAK1 mutant rice and in short-day poplar, it is obvious that K transporters do not transport Cs uniformly. Further analysis of the Cs transport capability in the individual transporters is required for identifying the substrate specificity of putative Cs transporters.

References

Alemán F, Caballero F, Ródenas R, Rivero RM, Martínez V, Rubio F (2014) The F130S point mutation in the Arabidopsis high-affinity K^+ transporter AtHAK5 increases K^+ over Na^+ and Cs^+ selectivity and confers Na^+ and Cs^+ tolerance to yeast under heterologous expression. Front Plant Sci 5:430

Furukawa J, Kanazawa M, Satoh S (2012) Dormancy-induced temporal up-regulation of root activity in calcium translocation to shoot in *Populus maximowiczii*. Plant Roots 6:10–18

Ishikawa S, Hayashi S, Abe T, Igura M, Kuramata M, Tanikawa H, Iino M, Saito T, Ono Y, Ishikawa T, Fujimura S, Goto A, Takagi H (2017) Low-cesium rice: mutation in *OsSOS2* reduces radiocesium in rice grains. Sci Rep 7:2432

Kurita Y, Baba K, Ohnishi M, Anegawa A, Shichijo C, Kosuge K, Fukaki H, Mimura T (2014) Establishment of a shortened annual cycle system; a tool for the analysis of annual re-translocation of phosphorus in the deciduous woody plant (*Populus alba* L.). J Plant Res 127:545–551

Nieves-Cordones M, Mohamed S, Tanoi K, Kobayashi NI, Takagi K, Vernet A, Guiderdoni E, Périn C, Sentenac H, Véry AA (2017) Production of low-Cs^+ rice plants by inactivation of the K^+ transporter OsHAK1 with the CRISPR-Cas system. Plant J 92:43–56

Nihei N (2016) Monitoring inspection for radioactive substances in agricultural, livestock, forest and fishery products in Fukushima prefecture. In: Nakanishi TM, Tanoi K (eds) Agricultural implications of the Fukushima nuclear accident: the first three years. Springer, Tokyo, pp 11–21

Noda Y, Furukawa J, Aohara T, Nihei N, Hirose A, Tanoi K, Nakanishi TM, Satoh S (2016) Short day length-induced decrease of cesium uptake without altering potassium uptake manner in poplar. Sci Rep 6:38360

Ohmori Y, Inui Y, Kajikawa M, Nakata A, Sotta N, Kasai K, Uraguchi S, Tanaka N, Nishida S, Hasegawa T, Sakamoto T, Kawara Y, Aizawa K, Fujita H, Li K, Sawaki N, Oda K, Futagoshi R, Tsusaka T, Takahashi S, Takano J, Wakuta S, Yoshinari A, Uehara M, Tanaka S, Nagano H, Miwa K, Aibara I, Ojima T, Ebana K, Ishikawa S, Sueyoshi K, Hasegawa H, Mimura T, Mimura M, Kobayashi NI, Furukawa J, Kobayashi D, Okochi T, Tanoi K, Fujiwara T (2014a)

Difference in cesium accumulation among rice cultivars grown in the paddy field in Fukushima prefecture in 2011 and 2012. J Plant Res 127:57–66

Ohmori Y, Kajikawa M, Nishida S, Tanaka N, Kobayashi NI, Tanoi K, Furukawa J, Fujiwara T (2014b) The effect of fertilization on cesium concentration of rice grown in a paddy field in Fukushima prefecture in 2011 and 2012. J Plant Res 127:67–71

Rai H, Yokoyama S, Satoh-Nagasawa N, Furukawa J, Nomi T, Ito Y, Fujimura S, Takahashi H, Suzuki R, Yousra E, Goto A, Fuji S, Nakamura S, Shinano T, Nagasawa N, Wabiko H, Hattori H (2017) Caesium uptake by rice roots largely depends upon a single gene, *HAK1*, which encodes a potassium transporter. Plant Cell Physiol 58:1486–1493

Technical Report Series No. 472 (2010) Handbook of parameter values for the prediction of radionuclide transfer in terrestrial and freshwater environments, vol 472. International Atomic Energy Agency, Vienna, pp 99–108

Part III
Radiocesium Movement Through Ecological Processes in Forest Ecosystem

Chapter 8
Movement of Radiocesium as Litterfall in Deciduous Forest

Tatsuhiro Ohkubo, Hiroko Suzuki, Mineaki Aizawa, and Kazuya Iizuka

Abstract Litterfall is an important agent in the process of radiocesium movement in forest ecosystems; however, the information is limited. We clarified the process of radiocesium movement in Konara oak forest and predicted the timing of resume of leaf litter for leaf litter origin compost production as management implications. During the initial stage after the FDNPP accident, forest canopy captured direct radiocesium fallout. However, deciduous forests acted less than evergreen conifer forest as the receptor of direct radiocesium fallout. Radiocesium concentration of litterfall and A_0 was exponentially decreased for 6 years. On the other hand, radiocesium concentration of A layer (0–5 cm) was increased. The organic A_0 layer is traditionally used for litter origin compost production. The availability of litterfall depended on the level of radiocesium contamination.

Keywords Konara oak · A_0 layer · Leaf litter origin compost · Radiocesium movement · Litterfall

8.1 Introduction

During the initial stages of radioactive cesium contamination after the Fukushima Daiichi Nuclear Power Plant (FDNPP) accident, forests, especially the canopies, acted as a receptor against direct radiocesium fallout and subsequently retained in forest ecosystems. In forest ecosystems litter is an important agent in the process of radiocesium movement from tree canopy to ground surface. Especially leaf litter on forest floor in deciduous broad-leaved forests in Satoyama, Japanese hilly landscape

---

T. Ohkubo (✉) · H. Suzuki · M. Aizawa
School of Agriculture, Utsunomiya University, Utsunomiya, Tochigi, Japan
e-mail: ohkubo@cc.utsunomiya-u.ac.jp

K. Iizuka
School of Agriculture, University Forests, Utsunomiya University, Shioya-gun, Tochigi, Japan

C. Takenaka et al. (eds.), *Radiocesium Dynamics in a Japanese Forest Ecosystem*,
https://doi.org/10.1007/978-981-13-8606-0_8

complex with paddy field and deciduous forests has been collected traditionally and used for leaf litter origin compost production for paddy rice, vegetable, and fruits production and also flowers growing. The leaf litter for compost production and direct application is severely contaminated by FDNPP accident. After the accident, the Ministry of Agriculture, Forestry and Fisheries formulated regulations after April 2012 for the use of forest products, setting permissible levels at 400 Bq kg^{-1} in biochar and leaf litter origin compost (MAFF 2011a, b). Most of commercial-based litter origin compost production was stopped by the accidents in the contaminated area. Toward the resume of commercial-based compost production, monitoring of radiocesium contamination in litterfall of the deciduous broad-leaved forest commenced since May 2011. The findings from the initial stages of the impact of FDNPP accident on the contamination in the litterfall are important to complete this information. And also a number of monitoring studies regarding radiocesium contamination in litterfall of conifer species started from the beginning of the accidents. The comparison of deciduous broad-leaved species with evergreen conifer species is also indispensable to clarify the pathways and processes within each forest ecosystem.

In this regard, we clarify the movement of radiocesium as litterfall especially in deciduous broad-leaved forests using litter traps, and the processes of radiocesium movement from living leaves and branches in the canopy are also analysed. Finally for the management implications, we tried to predict the timing of resume of leaf compost production comparing with permissible level of the compost use for food security.

8.2 Interception of Deposition by Tree Canopies

The FDNPP accident resulted in massive emission of radioactive substances into the atmosphere over a wide area; the only long-lived radionuclide released in significant amounts was ^{137}Cs (T/2 30.1 yr). Estimated total released was 19–24 PBq (Aoyama et al. 2016). During wet/dry deposition, radiocesium was intercepted by leaves/needles and branches of the canopy. The degree of interception is strongly controlled by the types of forest and the time of year (Shaw 2007). In case of the accident of the FDNPP, major release of radiocesium occurred at a stage prior to leaf flushing of deciduous trees. Therefore, a part of trees and forests of deciduous species (Konara oak, *Quercus serrata*) were much less contaminated, compared with evergreen conifer species (Sugi, Japanese cedar, *Cryptomeria japonica*) (Komatsu et al. 2016), e.g. cumulative deposition ratio of Konara oak to Sugi is 35–40% by Kato et al. (2017), and ^{137}Cs stock difference between Konara oak and sugi is 43% in highly contaminated area by Komatsu et al. (2016).

8.3 Temporal Migration of Radiocesium with Litterfall from the Canopy to the Forest Floor

Changes in radiocesium transfer to leaf litter are of serious concern with regard to the future use of leaf litter. Toward the resume of producing leaf litter origin compost at the commercial base, the monitoring of radiocesium contamination of fallen litter is indispensable. In order to clarify the change of radiocesium ($^{134}Cs + ^{137}Cs$) concentration in fallen leaves for 6 years (2011–2016) and to compare with that of leaf litter layer (A_0 layer) and surface soil (A layer, 0–5 cm), monitoring of radiocesium ($^{137}Cs + ^{134}Cs$) concentration in Konara-dominated forests since November 2011 (half year after the accident) started in three studied sites with different level of initial radiocesium contamination (high contamination area in Sekiya, intermediate contamination area in Funyu, and low contamination area in Ohgisu). The initial deposition of radiocesium fallout is calculated by third and fifth airborne monitoring by MEXT (2013). The values of high site in Sekiya, intermedium site in Funyu, and low site in Ohgisu are 147, 399 Bq/m^2, 37,027 Bq/m^2, and 4,489 Bq/m^2, respectively. Figure 8.1 shows the change of radiocesium ($^{134}Cs + ^{137}Cs$) concentration of fallen leaf litter (before and after leaf fall), A_0 layer, and A layer for 6 years (2011–2016) after the FDNPP accidents. The radiocesium concentration of the first year after the accident (2011 in Funyu and Ohgisu, 2012 in Sekiya) showed decreasing order of initial deposition, Sekiya, Funyu, and Ohgisu, respectively.

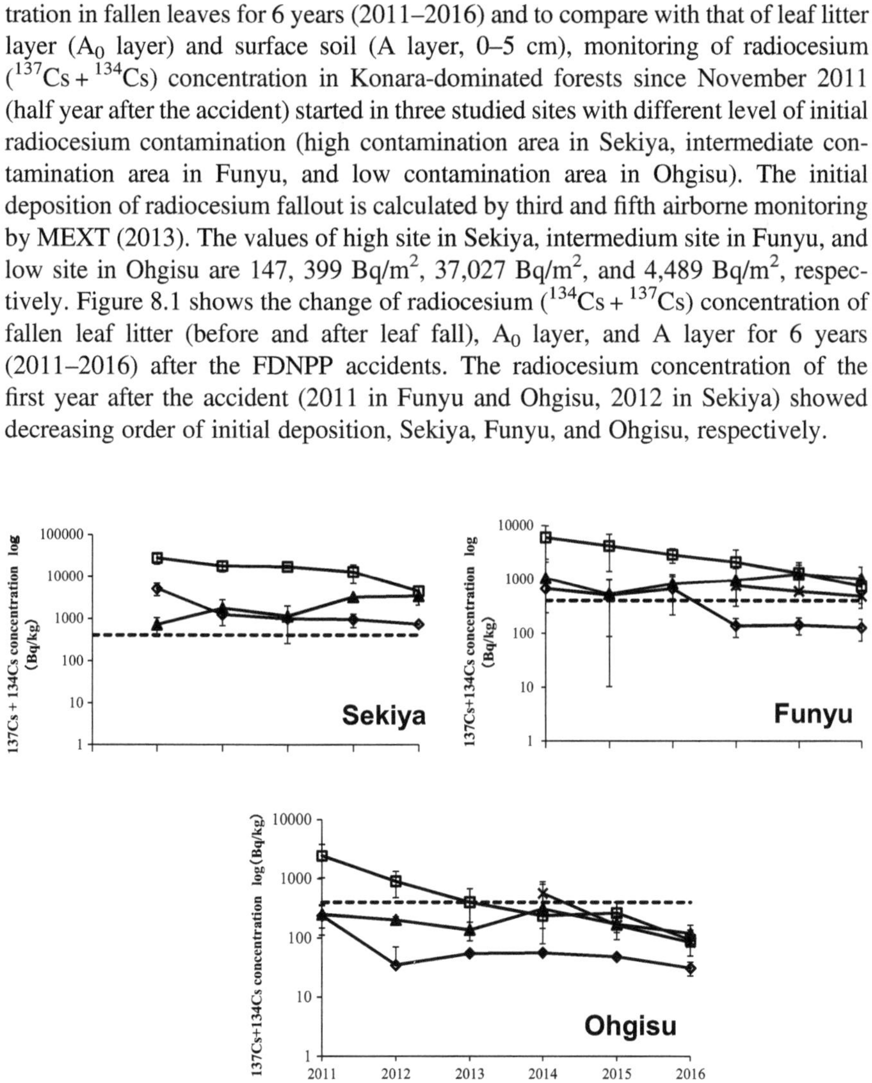

Fig. 8.1 Change of radiocesium ($^{134}Cs + ^{137}Cs$) concentration of fallen leaf litter (before and after leaf fall), A_0 layer, and A layer in three Konara oak forests (Sekiya, Funyu and Ohgisu) for 6 years after FDNPP accidents (2011–2016). x: litterfall (before fall), ◇:litterfall(after fall), □:A_0 layer, ▲:A layer (0-5 cm), -------: Permissible level (400 Bq/kg)

The reason why the differences of radiocesium contamination between fallen leaf litter and A_0 layer in the first year after the accident (2011) was derived from absence of leaves in the canopy before flushing in spring and collection of the litter in autumn, which brought the decrease of radiocesium concentration through wet deposition (rainfall, etc.) during the period. Radiocesium concentration of fallen leaves was exponentially decreased over time and much lower than that of leaf litter layer (A_0), and 5–6 years after the accident, the differences are getting smaller than the beginning. In low contamination site in Ohgisu, it showed lower radiocesium concentration than permissible level since the accidents in 2011 and subsequently slightly fluctuated.

On the other hand, the high and intermediate contamination sites in Sekiya and Funyu, it showed higher radiocesium concentration than permissible level in 2011 and 2012 and was gradually decreased after the peak. After 2014, it showed lower radiocesium concentration than permissible level in Funyu; however, it continued higher value than permissible level in Sekiya. After the year in 2014, radiocesium concentration of fallen leaf litter before autumn showed higher values, suggesting accumulation of radiocesium after that of the previous year. Radiocesium concentration in A (0–5 cm) layer gradually increased over time and became higher than that of A_0 layer. And radiocesium concentration in A (1–5 cm) layer in Ohgisu was stable over time. The above results showed that leaf litter and A_0 layer in Ohgisu could be used for producing compost after 2014 and after 2016 in Funyu but still could not be used in Sekiya.

8.4 Seasonal Changes of Radiocesium Concentration in Living Leaves and Branches in Canopy

Seasonal change of ^{137}Cs concentration of living leaves and branches in the Konara oak canopy and fallen leaves were monitored for 2 years in 2015 and 2016 (Fig. 8.2) in Sekiya. ^{137}Cs concentration of living leaves increased in spring of April and May, and slightly increased again in September, and decreased after the period in both years. For the ^{137}Cs concentration of living branches, that of current year, branches were higher than that of second and the older year ones and fluctuated similar to that of living leaves in canopy. ^{137}Cs concentration of old branches was lower than that of second year branches.

Although they had slightly seasonal fluctuations, living leaves of Konara oak showed high ^{137}Cs concentration (from 1,972 Bq/kg to 6,420 Bq/kg) in 5 years after the FDNPP accident in 2016. At the time of FDNPP accident in March 2011, as Konara did not flush new leaves, most of ^{137}Cs deposited on the forest floor. Thus, ^{137}Cs concentration of living leaves in the canopy was derived from the deposition on the tree surface, absorption, transfer, and accumulation of ^{137}Cs in branches and living leaves in the canopy. ^{137}Cs concentration reached high value of 6,420 Bq/kg in April 2016 and 6,087 Bq/kg in May 2016. During the periods, the flushed leaves have small amount of volume with high concentration, then

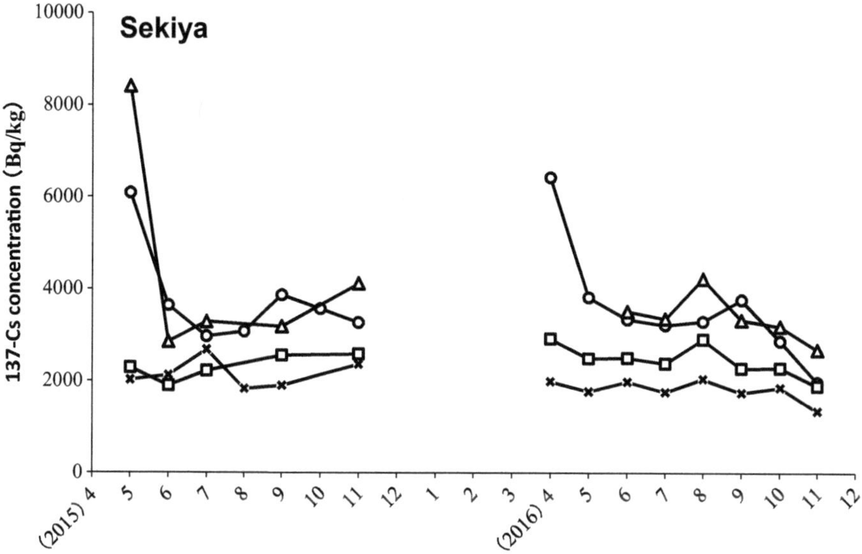

Fig. 8.2 Seasonal change of ^{137}Cs concentration of living leaves and branches in the Konara oak canopy and fallen leaves in Sekiya in 2015 and 2016. △: living leaves in canopy, □: ≧3 years old branches, ×: 2 years old branch, △: current year branch

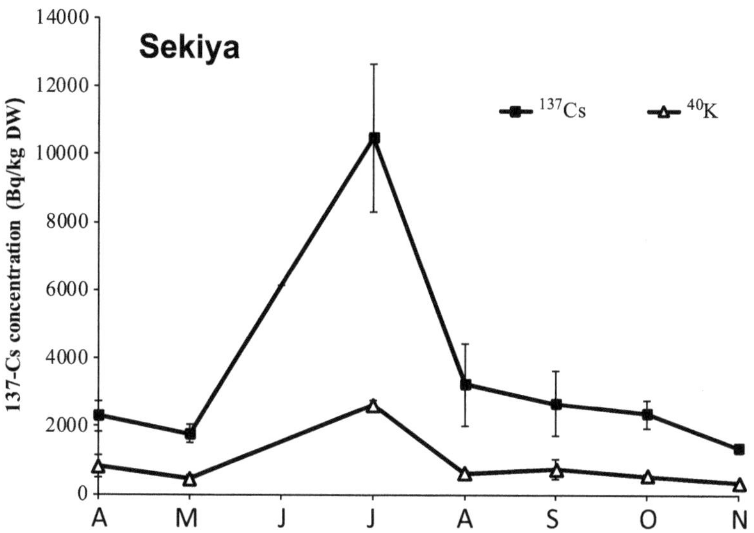

Fig. 8.3 Seasonal changes of ^{137}Cs and ^{40}K concentration in fallen litter in the Konara oak forest in Sekiya in 2016

decreasing the concentration of ^{137}Cs with increasing the amount of volume after June 2016 by dilution effect. Seasonal changes of ^{137}Cs and ^{40}K concentration in fallen litter in 2016 are shown in Fig. 8.3. Both values reached the highest value in

July, and before and after July, the values were getting lower. There are possible reasons why there was a peak of ^{137}Cs concentration in July. One is by rainfall, but there were less effects because of few heavy rain depositing radiocesium during July 2016. And the other possible reason is by litterfall of living leaves with high radiocesium concentration in April and May.

8.5 Possibility of Resume of Litter Origin Compost Production

For the comparison of radiocesium concentration in A_0 layer with the permissible level (400 Bq/kg) used for leaf origin compost, the concentration had decreased below the permissible level in 2013 in Ohgisu. On the other hand, in two sites in Sekiya and Funyu, the radiocesium concentration is still above the permissible level for the studied periods (2011, 2012–2016). Negative exponential equations were fitted to the decrease of radiocesium concentration in two sites, and the time of resume of litter origin compost production was predicted in 2019 in intermediate contamination site in Funyu and in 2023 in high contamination site in Sekiya (Fig. 8.4).

8.6 Studied Sites and Methods

Three studied sites with different level of radiocesium concentration were selected in a Konara oak (*Quercus serrata*) –pine (*Pinus densiflora*) forests in Sekiya, Nasushiobara city (N36 93′, E139 91′), in Konara oak forests in Funyu forest of Utsunomiya University Forests in Shioya town, (N36 79′, E139 82′), and Ohgisu in Nasukarasuyama city (N36 61′, E140 21′). The distances between FDNPP and the studied sites are about 112 km for Sekiya, about 129 km for Funyu, and about 113 km for Ohgisu, respectively. In the three sites, litterfall monitoring was started using litterfall traps (3 traps in Sekiya, 15 traps in Funyu, 3 traps in Ohgisu) made of nylon mesh net with the area of 0.5 m^2. Litter was collected in every month from April to November in 2016 in Sekiya and also collected at the times before and after litterfall (in September and December) in Funyu and Ohgisu in 2011–2016. The collected litter was sorted into leaves, branches, and others, and leaves were used for samples for analysis. Living branches with leaves were collected from canopies of five Konara oak trees in Sekiya in every month from April to November in 2016. Leaves were separated from branches, and the branches were cut into current, second, and old. Each leaf and branch was grounded using a mill (IFM-S10G; Iwatani, Tokyo, Japan) to prepare powder for measuring radiocesium concentration.

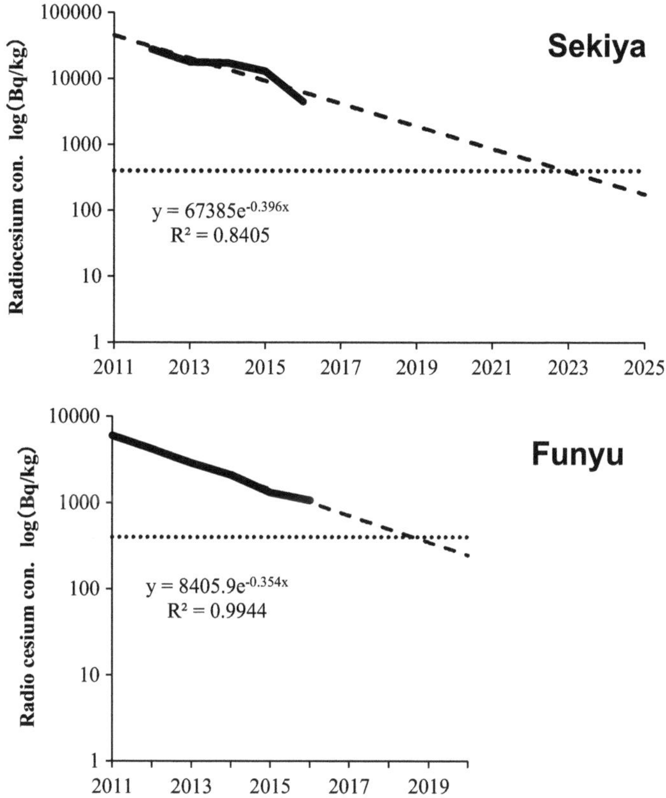

Fig. 8.4 Prediction of the time to resume of litter origin compost production using fitting negative exponential curves to the decrease of radiocesium concentration of A_0 layer in two sites (Sekiya and Funyu). ____: Radiocesium concentration in A_0 layer, -------: Permissible level (400 Bq/kg)

Radiocesium concentration for sample was determined by a gamma counter (Hitachi ARC-7000). ^{137}Cs (Bq/kg dry weight) concentration of sample was also determined with a germanium (Ge) semiconductor detector (Seiko EG & G, Ortec GEM30-70, Tokyo, Japan). Measurement conditions were as follows: measurement duration, 2000 sec or longer; gamma-ray peaks, 661.64 keV.

Acknowledgments This research was financially supported by JSPS KAKENHI under Grant numbers 24110001, 16K07769, 18K05738, 19K06140. We would like to thank many collaborators, Center for Bioscience Research and Education, and University Forest, Utsunomiya University, for their assistance in the radiocesium measurement and field work.

Compliance with Ethical Standards Conflict of interest. The authors declare that they have no conflict of interest.

References

Aoyama M, Kajino M, Tanaka TY, Sekiyama TT, Tsumune D, Tsubono T, Hamajima Y, Inomata Y, Gamo T (2016) ^{134}Cs and ^{137}Cs in the North Pacific Ocean derived from the March 2011 TEPCO Fukushima Dai-ichi nuclear power plant accident, Japan. Part two: estimation of ^{134}Cs and ^{137}Cs inventories in the North Pacific Ocean. J Oceanogr 72:67–76

Kato H, Onda Y, Hisadome K, Loffredo N, Kawamori A (2017) Temporal changes in radiocesium deposition in various forest stands following the Fukushima Dai-ichi Nuclear Power Plant accident. J Environ Radioact 166:449–457

Komatsu M, Kaneko S, Ohashi S, Kuroda K, Sano T, Ikeda S, Saito S, Kiyono Y, Tonosaki M, Miura S, Akama A, Kajimoto T, Takahashi M (2016) Characteristics of initial deposition and behavior of radiocesium in forest ecosystems of different locations and species affected by the Fukushima Daiichi Nuclear Power Plant accident. J Environ Radioact 161:2–10. https://doi.org/10.1016/j.jenvrad.2015.09.016

MAFF (2011a) About setting of provisional tolerance of radioactive cesium-containing fertilizer/soil improvement material/soil supplement and feed. http://www.maff.go.jp/j/syouan/soumu/saigai/shizai.html. Accessed 26 June 2017

MAFF (2011b) Q & A on setting immediate indicator values for firewood and charcoal for cooking and heating. http://www.rinya.maff.go.jp/j/tokuyou/shintan4.html. Accessed 26 June 2017

MEXT (2013) The third research on distribution of radioactive material derived from the Fukushima Daiichi Nuclear Power Plant accident, p 180 (in Japanese). http://fukushima.jaea.go.jp/initiatives/cat03/pdf05/03-2.pdf. Accessed 1 Dec 2018

Shaw G (2007) Radionuclides in forest ecosystems. In Shaw G (ed) Radioactivity in the terrestrial environment. Elsevier, Boston, 127–156pp

Chapter 9
Changes in Chemical Forms of Radiocesium in the Forest Floor Organic Matter with Decomposition and Uptake of Radiocesium Derived from the Organic Matter by Crops

Hitoshi Sekimoto

Abstract The behavior of radiocesium is affected by the availability of contaminated organic matter in the forest floor such as fallen leaves and humus. The forest floor organic matter contaminated with radiocesium would result in pollution of the farmland through application of organic matter as fertilizer. Then, firstly, changes in the chemical forms of radiocesium in the organic matter during decomposition were examined. In addition, decomposed organic matter was applied to the soil to evaluate the transfer of radiocesium to crops. Plant available radiocesium was considerably lower comparing with the total Cs radioactivity, with the ratio of plant available radiocesium to total radiocesium having shown little variation during the 18-month incubation. Plant available radiocesium ratio in fallen leaf compost was between 10% and 30% and about 1% in humus compost. The Cs derived from the compost by decomposition was captured in the soil. Therefore, as long as the organic matter containing radiocesium was in contact with the soil, the uptake of radiocesium from the organic matter by both komatsuna (*Brassica rapa*) and rice plant would be sufficiently suppressed. Even if the organic matters containing radiocesium at the concentration over the permission level were applied to farmland soil, the radiocesium concentrations in crops or rice would be remarkably lower than the permissible limit.

Keywords Forest floor organic matter · Plant available radiocesium · Radiocesium uptake by crops

H. Sekimoto (✉)
Scool of Agriculture, Utsunomiya University, Utsunomiya, Japan
e-mail: hitoshis@cc.utsunomiya-u.ac.jp

C. Takenaka et al. (eds.), *Radiocesium Dynamics in a Japanese Forest Ecosystem*,
https://doi.org/10.1007/978-981-13-8606-0_9

9.1 Introduction

The organic matters in the forest floor affected by deposition of radionuclides caused by the Fukushima Daiichi Nuclear Power Plant (FDNPP) accident contain radiocesium at relatively high level of concentrations. It is thought that the behavior of the radiocesium depends on the availability or unavailability of radiocesium in the contaminated organic matter of forest floor such as fallen leaves and humus, but it has not yet been confirmed. In some cases, the contaminated organic matter has been unexpectedly flowed into the environment or the adjacent farmland, or it has been artificially applied as a fertilizer into farmland. The radiocesium contamination in farmland as organic matters is a serious concern for farmers. Therefore, in order to clarify the chemical stability and plant availability of radiocesium in organic matter, the organic matters such as fallen leaves and humus were collected from the forest floor and decomposed experimentally. And the changes in the chemical forms of radiocesium in the organic matter during decomposition were examined. In addition, the decomposed organic matter was applied to soil to evaluate the transfer of radiocesium to crops.

9.2 Change in the Radiocesium Contained in Organic Matter with Decomposition

9.2.1 Change in the Content and Chemical Forms of Radiocesium in Organic Matter with Decomposition

The changes in the total and plant available radiocesium in fallen leaves and humus compost collected from the two forest sites (Sekiya and Takahara) in Tochigi prefecture were analyzed by the incubation experiment for 18 months. The moisture content of the composts changed by about 50–70% during incubation. The changes in total and plant available radiocesium are shown in Fig. 9.1. The total radiocesium in the compost made from fallen leaves increased by around 8.0 kBq/kg dry weight in Sekiya and 0.8 kBq/kg dry weight in Takahara. In contrast, the total radiocesium concentrations in the compost made from humus (A_0 layer) in both Sekiya and Mount Takahara were around 50 kBq/kg dry weight and 27 kBq/kg dry weight, respectively, and were significantly higher than that found in the compost made from fallen leaves. In general, because the weight of the compost decreases due to the decomposition of organic matter through the release of carbon dioxide and water, inorganic nutrients are concentrated in the compost. The gradual increase in total radiocesium observed in the compost suggests that the decomposition of fallen leaf and humus organic matter was at an advanced stage.

Plant available radiocesium was considerably lower than the total Cs radioactivity, with the ratio of plant available radiocesium to total radiocesium having shown

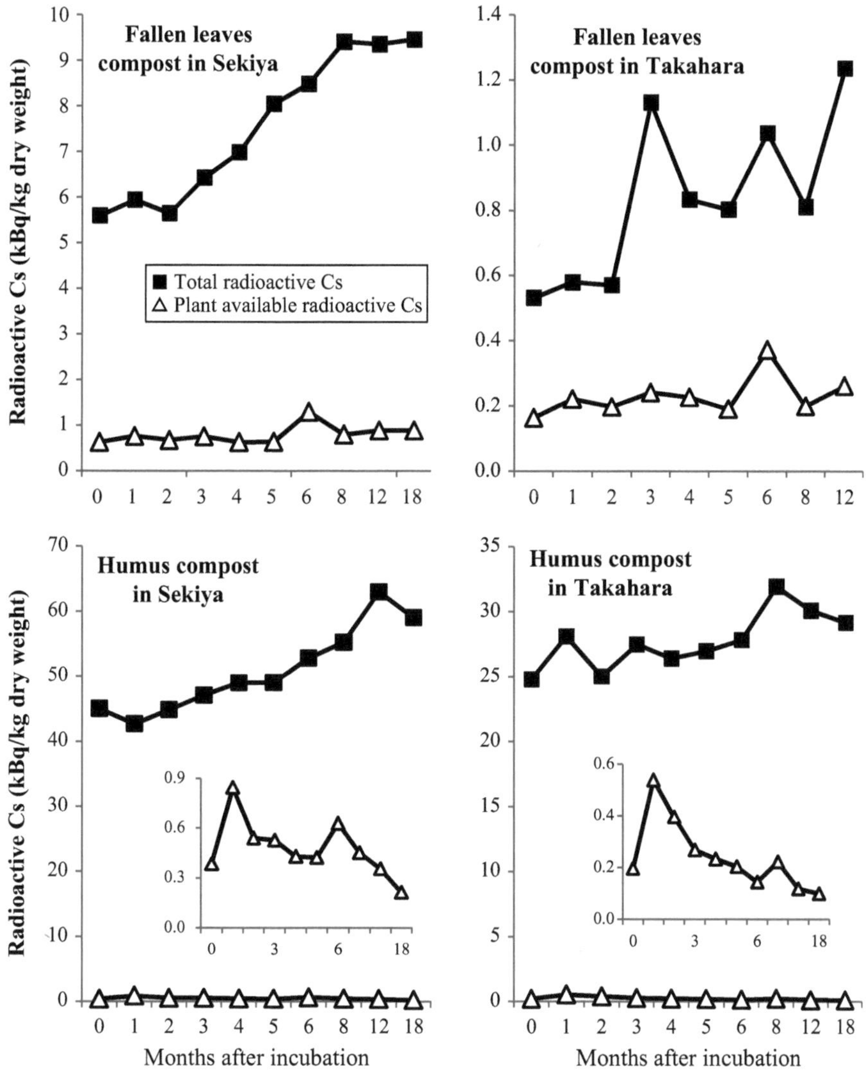

Fig. 9.1 Changes in total and plant available radiocesium in fallen leaves and humus compost with incubation

little variation during the 18-month incubation. Plant available radiocesium in compost was about 10% in Sekiya and 30% in Takahara for fallen leaf compost and about 1% in both Sekiya and Takahara for humus compost. Plant available radiocesium did not increase with the decomposition of the organic matter, so it was concluded that most of the radiocesium remained in the forest floor as a plant unavailable form.

9.2.2 Experimental Methods

9.2.2.1 Sampling and Incubation of the Forest Floor Organic Matter

Fallen leaves and humus (A_0 layer) were gathered from a mixed forest of konara (*Quercus serrata*) and Japanese red pine (*Pinus densiflora*) in Sekiya, Nasushiobara-shi, Tochigi Prefecture, Japan, and from a Japanese beech (*Fagus crenata*) forest near Mount Takahara, Shioya-machi, Tochigi Prefecture, on December 7, 2012. The radiocesium deposition flux immediately after the FDNPP accident (March 2011) had been measured as being in the range of 100 kBq/m^2 and 60 kBq/m^2 at the two locations, respectively. For 2 months from October to December 2012, a sheet was installed on the forest floor, and newly fallen leaves were collected. In addition, the litter layer under the sheet was removed and lower humus layer (A_0 layer) was sampled. The total Cs radioactivities (^{134}Cs+^{137}Cs) in the fallen leaves and humus were 8.8 and 60.4 kBq/kg dw in Sekiya 1.2 and 31.6 kBq/kg dw in Mount Takahara, respectively.

Under natural condition, fallen leaves decompose slowly on the forest floor, and the organic layer (humus) forms over the course of a year. In this study, fallen leaves and humus were decomposed under the following artificial conditions: the leaves and humus were each mixed with low-radioactivity rice bran (^{134}Cs +^{137}Cs:10.2 Bq/kg dry weight) as nitrogen source at a 10:1 volume ratio, incubated at 25 °C with 50–60% water by weight for 18 months from December 13, 2012, and were stirred once a week. Separate composts were made for the fallen leaves and the humus. A portion of each compost was sampled monthly.

9.2.2.2 Measurements of Radiocesium Contents in the Organic Matters

Samples of the compost were freeze-dried and powdered, and the dry powdered samples were stored in U-8 containers. The total radioactivity of Cs was measured by using a germanium semiconductor detector.

After the total radiocesium (^{134}Cs +^{137}Cs) of the dried sample of compost was measured, the availability of soluble radiocesium in the compost samples was carried out as follows: 100 mL of demineralized water was added to 30 g of dried powder sample, and the mixture was shaken for 1 h. The extract solution was filtered through a 1.0 μm glass filter followed by a 0.45 μm membrane filter. This extraction was repeated three times, with a total of 300 mL of extract containing water-soluble radiocesium obtained. The sample was then further extracted three times using 100 mL of 0.05 M hydrochloric acid, and 300 mL of extract containing ion-exchangeable radiocesium was obtained. Usually, the ion-exchangeable fraction is extracted using ammonium acetate, however, because it was necessary to evaluate the fertilizer efficiency of the extracted residue, extraction was undertaken using hydrochloric acid instead to avoid the use of ammonia (Harada and Inoko 1980). This process is shown in Fig. 9.2.

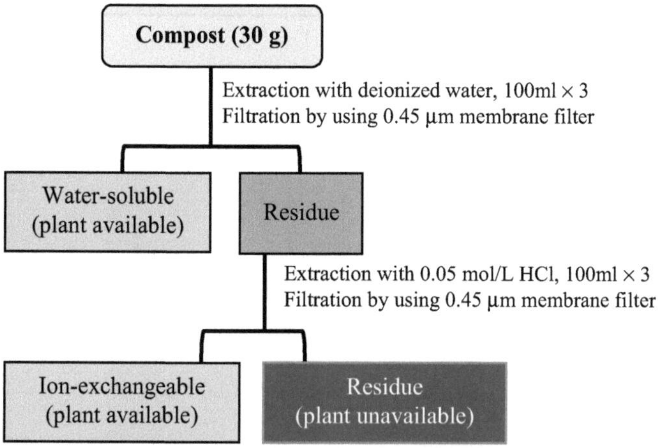

Fig. 9.2 Preparation of water-soluble and ion-exchangeable (plant available) extracts

Subsequently, 100 mL of each extract was placed in a U-8 container, and radiocesium was measured. The sum of water-soluble and ion-exchangeable radiocesium was defined as plant available radiocesium. Consequently, the final residue contained Cs that was unavailable to plants. After a part of the final residue was washed with demineralized water, an experiment was conducted to evaluate uptake of radiocesium by crops that had this residue applied to them (see next Sect. 9.3).

9.3 Radiocesium Uptake by Crops from the Radiocesium-Contaminated Organic Matter

*9.3.1 Radiocesium Uptake by Komatsuna (*Brassica rapa*)*

The uptake of radiocesium from contaminated organic matter by crops is affected by the chemical form of radiocesium, such as plant available form or unavailable form. The pot experiment using four kinds of contaminated composts, which are the litter compost with extraction of plant available radiocesium, the litter compost without extraction, the humus compost with extraction, and the humus compost without extraction, was conducted for the growth of komatsuna (*Brassica rapa*). There was no difference in growth among the experimental pots. The radiocesium of komatsuna (*Brassica rapa*) was increased by the application of compost containing radiocesium. In particular, radioactivity in the plants tended to increase by the compost with extraction of plant available radiocesium (Fig. 9.3).

The pH of compost before and after extraction was measured and was found to be around pH 5.8 for both composts, suggesting that the remaining dilute hydrochloric acid in the compost after the extraction of plant available radiocesium had no influence. In contrast, the concentration of K contained in composts before and

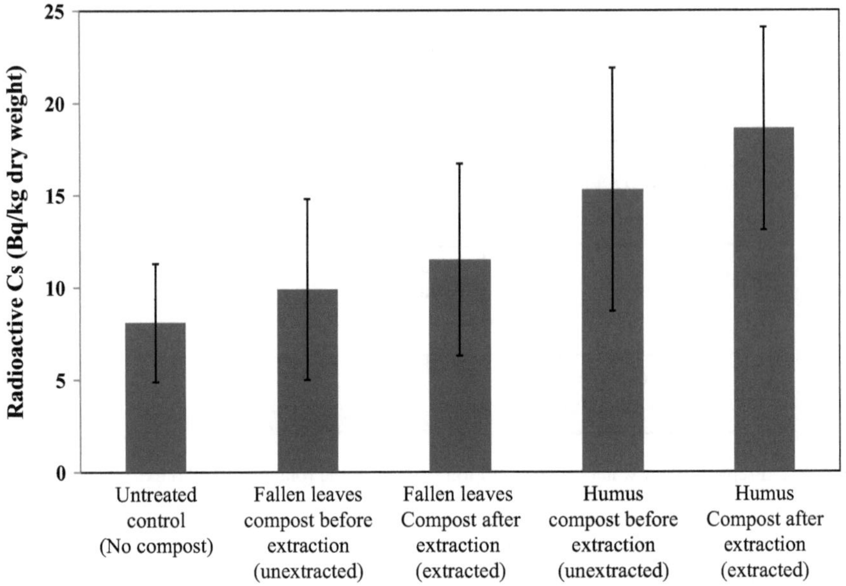

Fig. 9.3 Radiocesium of *Brassica rapa* in different compost treatments. Error bars are standard deviations ($n = 4$)

after extraction was confirmed to have decreased. The K concentration was 298 mmol/kg in litter compost (before), 31 mmol/kg (after), and 170 mmol/kg in humus compost (before) and 31 mmol/kg (after), respectively. The K concentration of composts after extraction was clearly low. Therefore, it was considered that the absorption of radiocesium would be promoted in the compost after extraction by the low K concentration.

Nonetheless, if organic matter containing high radiocesium at a level that is not used in practice should be applied in contact with the soil, the radiocesium of *Brassica rapa* would be remarkably low, as the observed radiocesium of *Brassica rapa* was lower than the permissible limit.

9.3.2 Radiocesium Uptake by Rice

There was no difference in growth and yield between experimental plots. The radiocesium of each rice plant is shown in Table 9.1. For the rhizosphere application treatment, the radiocesium of straw, chaff, and brown rice increased. This was due to direct contact of the compost containing radiocesium with the plant roots. In contrast, the radiocesium of rice plants in the soil application treatment was the same as that of the untreated control, i.e., the radiocesium uptake from the soil did not occur. It was concluded that radiocesium derived from the compost was captured

Table 9.1 Radiocesium concentrations in rice plant by different compost treatments

		Radioactive Cs of rice plant (Bq/kg dry weight)		
Compost	Application	Straw	Chaff	Brown rice
Fallen leaves	Rhizosphere	23.6 ± 1.9	13.7 ± 2.5	5.9 ± 0.8
	Soil	7.9 ± 2.2	7.1 ± 3.0	2.9 ± 0.2
Humus	Rhizosphere	70.5 ± 18.0	53.8 ± 8.8	14.3 ± 3.9
	Soil	11.7 ± 3.0	13.4 ± 4.6	3.1 ± 1.1
Untreated control		8.0 ± 1.1	6.8 ± 5.8	2.4 ± 1.1

Average ± SD ($n = 3$)

in the soil, and therefore, as long as the organic matter containing radiocesium in the forest floor was in contact with the soil, the uptake of radiocesium from the organic matter in rice plant will be sufficiently suppressed.

9.3.3 Experimental Methods

9.3.3.1 Radiocesium Uptake by Crops

Six-month-incubated composts of fallen leaves and humus from Sekiya, either unextracted or extracted following extraction of plant available radiocesium, were applied to common paddy field soil ($^{134}Cs + ^{137}Cs$:130.1 Bq/kg dry weight) and upland soil ($^{134}Cs + ^{137}Cs$:93.0 Bq/kg dry weight). However, the permissible limit of radiocesium in soil improvement materials such as compost is 400 Bq/kg (product weight), so the composts used in this research exceeded this level, and they could not be used in practice. Rice (*Oryza sativa*) and komatsuna (*Brassica rapa*) were cultivated to investigate the transfer of radiocesium to crops.

According to Aizawa et al. (2013), fallen leaf compost of 3.73 Mg dry weight/ha is required for rice cultivation. For the rice study, 10 g of compost was added to each experimental pot filled with 4 kg of paddy soil. For komatsuna cultivation, 5.0 g of compost was applied to 1 kg of upland soil.

9.3.3.2 Radiocesium Uptake by Leaf Vegetable, Komatsuna

Komatsuna (*Brassica rapa*) was sown into 1 L pots filled with 800 g of upland soil that had composts and 2.4 g of NPK compound fertilizer (10-18-16) applied. The compost applied were fallen leaves compost and humus compost. The compost added was either unextracted or extracted with water and acid (after extraction of plant available radiocesium). An untreated control contained no compost. Shoots of *Brassica rapa* were harvested 40 days after sowing, prepared as dry powder and used to measure radiocesium.

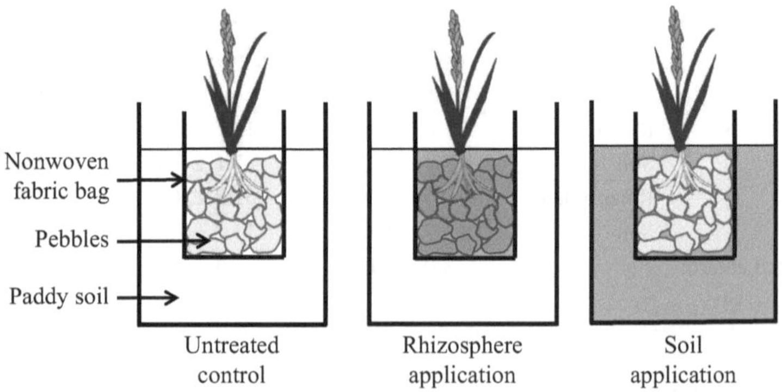

Fig. 9.4 Experimental setup to test the uptake of radiocesium from contaminated forest floor organic matter by rice

9.3.3.3 Radiocesium Uptake by Rice

A one-L nonwoven fabric bag filled with 1 kg pebbles was placed inside a 4 L pot, and 2.5 kg of paddy soil was packed around it (Fig. 9.4). The nonwoven fabric bag prevented the roots of rice transplanted inside it from intruding into the surrounding soil. Then, 8 g of NPK compound fertilizer (10-18-16) was applied to the soil component in the pot. Three experimental plots were used: the compost unextracted was applied either to the pebbles in the nonwoven fabric bag (rhizosphere application treatment) or to the soil outside of the bag (soil application treatment) and an untreated control. After harvesting, each rice plant was separated into straw, chaff, and brown rice. This was prepared as a dry powder, and the radiocesium was measured.

References

Aizawa M, Inui T, Hirai H, Osawa K, Ikeda J, Ohkubo T (2013) Yields of fallen leaves and the compost from a farm forest and their efficacy for paddy rice cultivation in a hilly and lower mountainous area in northern Kanto. Jpn J For Environ 55:119–126. (in Japanese with English abstract)

Harada Y, Inoko A (1980) The measurement of the cation-exchange capacity of composts for the estimation of the degree of maturity. Soil Sci Plant Nutr 26:127–134

Chapter 10
Contamination and Transfer of Radiocesium in Soil Ecosystem

Nobuhiro Kaneko

Abstract The fate of radio-Cs contamination in forest ecosystem after the Fukushima Daiichi Nuclear Power Plant accident was reviewed. First the fallout contaminated aboveground part of the forest, and then most of the contaminants moved downward in the forest and finally accumulated in the very surface layer of soil. Although any acute effect of radiation to forest organisms was not observed, soil organisms were contaminated via habitat and food. Feeding habit of soil organisms well reflected species-specific time-course changes in body contamination of radio-Cs. Mycoextraction of radio-Cs is promising method to rehabilitate contaminated forest floor using wood chips.

Keywords Decontamination · Forest products · Soil ecosystem · Soil food web

10.1 Introduction

The nuclear accident of Fukushima Daiichi Nuclear Power Plants (FDNPP) after Great East Japan Disaster released a vast amount of artificial nuclides, among which radiocesium was a major pollutant of this accident and having the relatively long half-life; thus, the environmental effect of radiation by radiocesium is a major concern of this accident. The soil is known to be a sink of radiocesium which polluted terrestrial ecosystems via fallout originating from anthropogenic sources. Long-term effects and decontamination of radiocesium are thus the major problems on natural resource use in the contaminated area.

Decontamination by removal of contaminated soil and potassium fertilization is effective countermeasure of croplands, whereas these actions are not useful in forest soil, because of the landscape, the extent of the total target area, and the

N. Kaneko (✉)
Faculty of Food and Agricultural Sciences, Fukushima University, Fukushima, Japan
e-mail: kaneko-nobuhirro@agri.fukushima-u.ac.jp

C. Takenaka et al. (eds.), *Radiocesium Dynamics in a Japanese Forest Ecosystem*,
https://doi.org/10.1007/978-981-13-8606-0_10

different soil management in the forest. Adding to that, there are fewer residents in the forest area, and economic activities are smaller compared to the urban and agricultural areas. The report by IAEA to the Japanese government on rehabilitation of contaminated forests by the Fukushima Accident (IAEA 2013; Ministry of Environment 2014) concluded that forest decontamination is not recommended because of low efficiency in reducing radiation risk to the public. Therefore, the incentive to conduct decontamination of forest soil is very limited, and the Japanese government officially declared that forest decontamination is out of focus in countermeasure of the pollution.

People used to use forest products and its ecosystem services, and the forests contaminated by radiocesium came from the power plants is not an exception. Besides wood timber, a major economic value, wild plants, animals, and especially edible mushrooms have been used by local residents. The contamination of these forest products is thoroughly explained in Part II. Among the forest products, a log for mushroom culture medium had been intensively utilized in the area of Fukushima Prefecture. Especially a log for shiitake mushroom media produced in Fukushima Prefecture has been highly evaluated and dominated Japanese market of a log for mushroom medium. As shown in Part II, the combination of konara oak as medium and cultivating shiitake mushroom for food production leads to higher contamination of radiocesium to shiitake mushroom. Therefore, harvesting the logs was inevitably stopped after the accident.

Soil ecosystem contains most terrestrial organisms and is supporting the aboveground ecosystem (Bardgett and van der Putten 2014). Biological interactions between plant root, microorganisms, and soil animals are driving plant growth by promoting nutrient cycling and soil structural modification (Kulmatiski et al. 2014). Soil animals are also linking below- and aboveground ecosystem, because they are a major prey for the aboveground animals (Chen and Wise 1999). It has been well-known that radiocesium accumulates in the very surface layer of mineral soil (Monna et al. 2009; Rafferty et al. 2000), and it was also confirmed after the Fukushima Accident in agricultural (Yamaguchi et al. 2016) and forest soils (Kato et al. 2012). Therefore, contamination and transfer of radiocesium in the soil layer need to be studied to understand the fate of radiocesium in the terrestrial ecosystem, root uptake of radiocesium from the soil, biomagnification of radiocesium by soil animals, and radiocesium transfer from soil to aboveground ecosystem via food web.

Detailed monitoring of the radiocesium migration and redistribution is fundamental. Here I will review the recent studies on radiocesium dynamics in soil ecosystem during the early phase of contamination and show our results on radiocesium dynamics in soil organisms.

There are strong demands by local residents to use forest services contaminated by radiocesium. It is essential to conduct countermeasures to mitigate the impacts of radioactive contamination of forests. I propose a "mycoextraction" as a potential decontamination method.

10.2 Contamination and Movement from Canopy to Forest Floor

The radiocesium contaminated both croplands and forests. The estimated total extent of heavily contaminated forest (>=^{134}Cs, ^{137}Cs 1000 kBq m^2) was 428 km^2, which is equivalent to 4% of the forests in Fukushima Prefecture (Hashimoto et al. 2012).

Previous studies have shown that ^{137}Cs transfer via throughfall from aboveground to belowground contributed more than those of defoliation in the very early phase after the fallout (Loffredo et al. 2014), and then the litter fall became a major translocation route of ^{137}Cs from canopy to forest floor (Teramage et al. 2014; Kato et al. 2017).

Imamura et al. (2017) examined four different vegetation types (red pine, Japanese cedar, hinoki cypress, and konara oak) in five locations with different contamination levels of radiocesium and found that the changes in ^{137}Cs in needles/leaves and wood differed among tree species, whereas the pattern of temporal changes (2011–2015) in radiocesium in the total of nine forest plots was similar overall. The ^{137}Cs in the organic layer decreased, and those in the 0–5 cm soil increased, and litter migration into deeper soil was observed.

According to the modeling study by Hashimoto et al. (2013), radiocesium in both canopy and organic soil layer decreased rapidly, whereas radiocesium in mineral soil layer shows peaks in the second (broadleaved forest) and the third year (coniferous forest) after the fallout, and most contaminants accumulate in the mineral layer.

10.3 Acute Effect and Long-Term Exposure in Soil Ecosystem

Zaitsev et al. (2014) reviewed the acute effect of ionizing radiation effects on soil animals. There are substantial studies on the effects of environmental pollution by radionuclides in the former Soviet Union territory (Geras'kin et al. 2007). The radiation level of 1 Gy/day caused reduction of soil invertebrate population density into half of the natural population density (Zaitsev et al. 2014).

Reduction of population density and activities of soil animals may retard organic matter decomposition. Mousseau et al. (2014) found a reduction in decomposition rate of forest litter around Chernobyl NPP.

10.4 Biomagnification and Transfer of Radiocesium Through a Food Chain

Calmon et al. (2009) reported transfer efficiency of radionuclides on various wild organisms. Mushrooms tend to show the highest TF value among the organisms examined, and some mushroom species showed biomagnification of radiocesium.

Table 10.1 The estimated initial contamination of radiocesium

Site ID	^{137}Cs (kBq/m^2)[a]
FA	1097
FB	404
FC	838
FD	229
FE	2455
FF	2542
FG	2037
FH	216
FI	5248
FJ	4981

[a]Estimated values on 2 July 2011

Murakami et al. (2014) measured radiocesium concentration of major forest organisms after FDNPS accidents. They found that there was no obvious biomagnification of radiocesium; however, detritus on the forest floor was a major source of the contaminants on the forest food web; detritus feeders and fungi showed the higher concentration of radiocesium compared to the herbivores and predators.

We studied changes in the radiocesium concentration of three soil organisms in ten Japanese cedar stands between 2011 and 2015 (see Method). The estimated initial contamination of radiocesium ranged from 219 to 4981 kBq/m^2 (Table 10.1). Irrespective to the initial contamination level, radiocesium concentration in the litter layer decreased, whereas those in the soil layers (0–5 cm) increased, thus radiocesium migrated downward from the organic to the mineral soil layer from 2011 to 2015 (Fig. 10.1).

We collected earthworms from 2 and 1 sites from the ten sites in 2011 and 2015, respectively, and found four species. All these species were epigeic: surface living species feeding litter with mineral soil. Radiocesium concentrations of earthworm (excluded gut contents) were higher than the litter radiocesium in one stand and lower in another stand, whereas body radiocesium was higher than soil in 2011 (Fig. 10.2). In 2015, body radiocesium concentration was lower than litter and soil.

A terrestrial Isopoda, *Ligidium japonica*, known to distribute wide, are in the Japanese forest. This species feeds on dead leaves; there was no clear relationship observed between body *L. japonica* (without gut contents) and litter layer in 2011, whereas there were significant positive correlations between the body with soil in 2011 and with litter and soil in 2015 (Fig. 10.3).

A saprotrophic fungus, *Strobilurus ohshimae*, produce fungal bodies in autumn, is often found in fallen twigs of Japanese red cedar. There were no clear relationships between environmental and fungal body contaminations in 2011, whereas there were significant positive relations in 2015 (Fig.10.4).

The transfer factors were decreased in the earthworm (one site) but increased in the Isopoda (Table 10.2). The fungus shown decreases in the litter, whereas it increases in the soil. The tag was similar in the earthworm and increased in the Isopoda, whereas it was decreased in the fungus. These different changes between

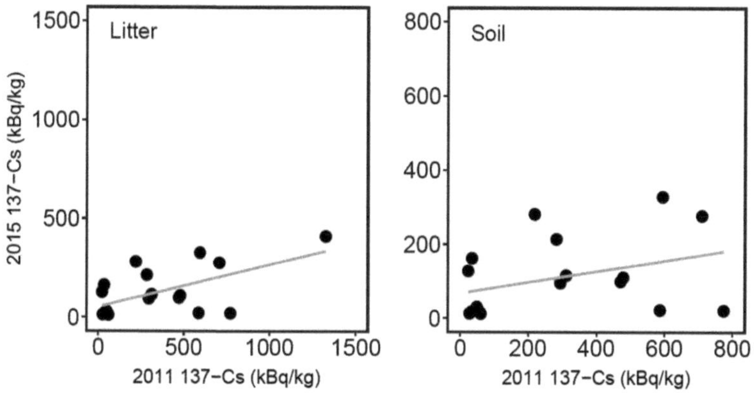

Fig. 10.1 Comparison of ^{137}Cs concentration between 2011 and 2015 of litter and soil at ten study stands

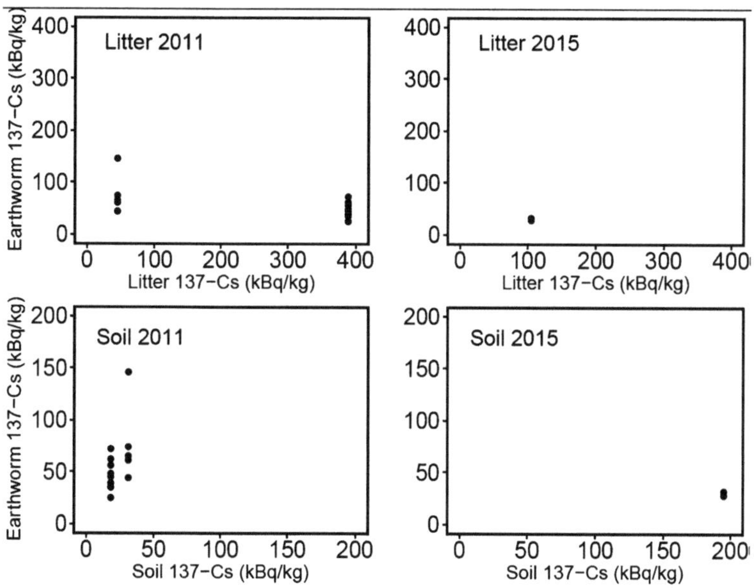

Fig. 10.2 Comparison of ^{137}Cs concentration between 2011 and 2015 of earthworms

organisms reflect feeding habits; the earthworms found in this study are litter feeders. The Isopoda are also litter feeder, but they tend to prefer aged litter and probably feed on organic matter in fermentation layer, therefore, radiocesium migration from fresh to aged litter and fermentation layer, where aged litter accumulates affected the Isopoda contamination. The fungus utilized fresh twigs on the forest floor. Contamination of leaves and twigs in Japanese red cedar decreased in 2015; thus, fungal contamination also decreased. Knowing radiocesium dynamics in the forest soil environment is useful in understanding the contamination level of radiocesium.

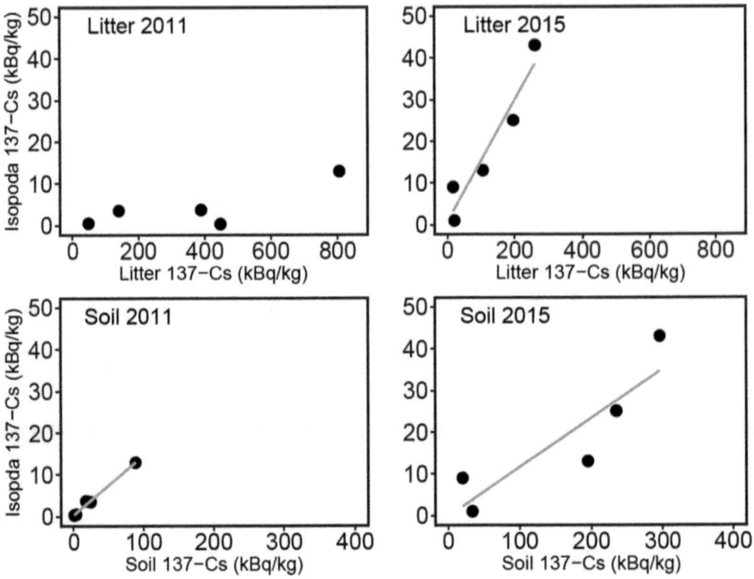

Fig. 10.3 Comparison of ^{137}Cs concentration between 2011 and 2015 of terrestrial isopod, *Ligidium japonicum*

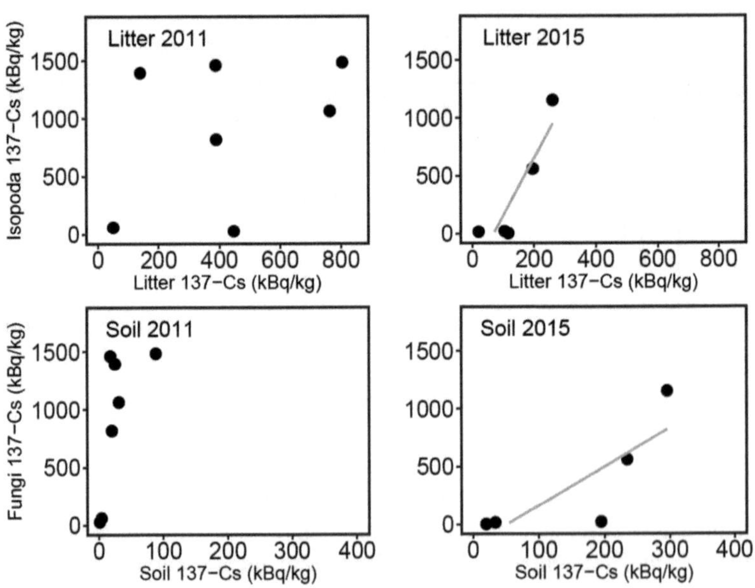

Fig. 10.4 Comparison of ^{137}Cs concentration between 2011 and 2015 of fungus, *Strobilurus ohshimae*

Table 10.2 Transfer factors of three soil organisms

Taxon	Year	TF litter	TF soil	Tag
Earthworms[a]				
	2011	0.319	2.442	0.052
	2015	0.285	0.154	0.041
Isopoda				
	2011	0.0007–0.0393	0.0014–0.03802	0.0011–0.0073
	2015	0.072–0.563	0.005–0.017	0.0049–0.0148
Fungus				
	2011	15.250–81.167	0.063–9.964	0.029–1.682
	2015	0.016–4.427	0.095–3.889	0.001–0.355

[a]One site where earthworms were collected in both 2011 and 2015

10.5 Countermeasure

Countermeasures to rehabilitate forest ecosystem is a big challenge due to its landscape and complex species interactions. Even after knowing that most contaminant stays in the very surface of the soil layer, soil replacement in a forest is not feasible when considering the treatment of removed soil.

Phytoremediation has been tested to decontaminate radiocesium, and it has been known that the method is not so effective to apply croplands compared to soil scraping and conversion tillage (Yamaguchi et al. 2016).

We have reported that fungal immobilization of radiocesium from soil to litter during the early phase of decomposition on the forest floor (Huang et al. 2016). The amount of radiocesium translocated from soil to litter after 1 year of the start of decomposition was estimated to be 4% (Huang et al. 2016). In that study, we used tree leaf litter that was equivalent to 256 g/m^2, which is around minimum annual leaf litterfall in deciduous forests. If we can increase the amount of organic matter as a growing medium for fungi on the forest floor, then we can expect to increase the amount of radiocesium translocation from soil to the medium.

Wood is the most abundant biomass in the forest, and also wood-decaying fungi are major decomposers in the forest ecosystem. Nutrient element concentrations in woods are known to be lower compared to leaves; therefore, fungi growing on woods need to get nutrients out from woods to keep the stoichiometric balance between essential nutrients. We presumed that putting a large number of wood chips on the forest floor at one time might stimulate fungal growth, and it will lead mass translocation of nutrients including radiocesium from soil to the wood chips. The amount of translocation will be determined by various factors: environmental conditions and duration, amount and shape of wood chips, contamination level, the presence of leaf litter, and trait of fungal species growing on the chips. Our results indicated that absorption to chips was estimated from 6.3% to 12.2% and 6.0 to 17.8% of total contaminants in the soil after 11 and 23 months on the soil surface (Kaneko et al. 2015).

It is not clear what kind of microorganisms is responsible for the translocation of radiocesium. AM fungal hyphae facilitated transfer among forbs and grasses (Meding and Zasoski 2008). Cs and Rb were used as K analogs and showed similar movement. There were many observations of radiocesium increase during the early phase of litter decomposition (Fukuyama and Takenaka 2004; Bruckmann and Wolters 1994).

Biological method to decontaminate radiocesium from soil is promising because there will be less disturbance and cost compared to engineering methods (e.g., soil washing). Mycoexraction is not as effective compared to soil scraping. However, this method can promote forest management; especially coppice rotation in the konara log production for the mushroom medium has been stopped after the accident. Appropriate management method is highly needed to continue forest management in these forests.

10.6 Method

We selected ten stands of Japanese red cedar (*Cryptomeria japonica*) plantations in the Abukuma Mountains with various contamination levels. A 10 × 10 m plot was set in each stand for a sampling of contaminated litter, soil, and organisms. A sampling of litter (50 × 50 cm), soil (0–5 cm), and soil organisms were conducted on 13 to 16 October 2011 and 19 to 22 September 2015. Earthworms and terrestrial isopods (*Ligidium japonicum*) were collected by hand sorting from 50 × 50 cm area down to 30 cm. Fungus (*Strobilurus ohshimae*) was collected from fallen twigs within the 10 × 10 m area.

Radiocesium concentration was determined by direct gamma-spectrometry using a Ge-detector for 600 s (CANBERRA GC2018). Radiocesium concentrations were shown after calculating the values at the time of field sampling considering radioactive decay.

References

Bardgett RD, van der Putten WH (2014) Belowground biodiversity and ecosystem functioning. Nature 515:505–511. https://doi.org/10.1038/nature13855

Bruckmann A, Wolters V (1994) Microbial immobilization and recycling of Cs-137 in the organic layers of forest ecosystems: relationship to environmental conditions, humification, and invertebrate activity. Sci Total Environ 157:249–256. https://doi.org/10.1016/0048-9697(94)04288-x

Calmon P, Thiry Y, Zibold G et al (2009) Transfer parameter values in temperate forest ecosystems: a review. J Environ Radioact 100:757–766. https://doi.org/10.1016/j.jenvrad.2008.11.005

Chen B, Wise D (1999) Bottom-up limitation of predaceous arthropods in a detritus-based terrestrial food web. Ecology 80:761–772

Fukuyama T, Takenaka C (2004) Upward mobilization of 137Cs in surface soils of *Chamaecyparis obtusa* Sieb. et Zucc. (hinoki) plantation in Japan. Sci Total Environ 318:187–195. https://doi.org/10.1016/s0048-9697(03)00366-8

Geras'kin SA, Evseeva TI, Belykh ES et al (2007) Effects on non-human species inhabiting areas with enhanced level of natural radioactivity in the north of Russia: a review. J Env Radioact 94:151–182. https://doi.org/10.1016/j.jenvrad.2007.01.003

Hashimoto S, Ugawa S, Nanko K, Shichi K (2012) The total amounts of radioactively contaminated materials in forests in Fukushima, Japan. Sci Rep 2:416. https://doi.org/10.1038/srep00416

Hashimoto S, Matsuura T, Nanko K et al (2013) Predicted spatiotemporal dynamics of radiocesium deposited onto forests following the Fukushima nuclear accident. Sci Rep 3:2564. https://doi.org/10.1038/srep02564

Huang Y, Kaneko N, Nakamori T, Miura M, Tanaka Y, Nonaka M, Takenaka C (2016) Radiocesium immobilization to leaf litter by fungi during first-year decomposition in a deciduous forest in Fukushima. J Environ Radioact 152:28–34

Imamura N, Komatsu M, Ohashi S et al (2017) Temporal changes in the radiocesium distribution in forests over the five years after the Fukushima Daiichi nuclear power plant accident. Sci Rep 7:8179. https://doi.org/10.1038/s41598-017-08261-x

International Atomic Energy Agency (2013). http://www.env.go.jp/press/files/jp/23189.pdf

Kato H, Onda Y, Teramage M (2012) Depth distribution of 137Cs, 134Cs, and 131I in soil profile after Fukushima Dai-ichi nuclear power plant accident. J Environ Radioact 111:59–64. https://doi.org/10.1016/j.jenvrad.2011.10.003

Kaneko N, Huang Y, TNakamori T (2015) Decontamination of Radio Cs from Forest Soils Using Biodiversity and the Functioning of Soil Organisms. J Japanese Forest Soc 97 (1):75–80

Kato H, Onda Y, Hisadome K et al (2017) Temporal changes in radiocesium deposition in various forest stands following the Fukushima Dai-ichi nuclear power plant accident. J Environ Radioact 166:449–457. https://doi.org/10.1016/j.jenvrad.2015.04.016

Kulmatiski A, Anderson-Smith A, Beard KH et al (2014) Most soil trophic guilds increase plant growth: a meta-analytical review. Oikos 123:1409–1419. https://doi.org/10.1111/oik.01767

Loffredo N, Onda Y, Kawamori A, Kato H (2014) Modeling of leachable 137Cs in throughfall and stemflow for Japanese forest canopies after Fukushima Daiichi nuclear power plant accident. Sci Total Environ 493:701–707. https://doi.org/10.1016/j.scitotenv.2014.06.059

Meding SM, Zasoski RJ (2008) Hyphal-mediated transfer of nitrate, arsenic, cesium, rubidium, and strontium between arbuscular mycorrhizal forbs and grasses from a California oak woodland. Soil Biol Biochem 40:126–134. https://doi.org/10.1016/j.soilbio.2007.07.019

Monna F, van Oort F, Hubert P et al (2009) Modeling of (137)Cs migration in soils using an 80-year soil archive: role of fertilizers and agricultural amendments. J Environ Radioact 100:9–16. https://doi.org/10.1016/j.jenvrad.2008.09.009

Mousseau TA, Milinevsky G, Kenney-Hunt J, Møller AP (2014) Highly reduced mass loss rates and increased litter layer in radioactively contaminated areas. Oecologia. https://doi.org/10.1007/s00442-014-2908-8

Murakami M, Ohte N, Suzuki T et al (2014) Biological proliferation of cesium-137 through the detrital food chain in a forest ecosystem in Japan. Sci Rep 4:3599. https://doi.org/10.1038/srep03599

Rafferty B, Brennan M, Dawson D, Dowding D (2000) Mechanisms of Cs-137 migration in coniferous forest soils. J Environ Radioact 48:131–143. https://doi.org/10.1016/s0265-931x(99)00027-2

Teramage MT, Onda Y, Kato H, Gomi T (2014) The role of litterfall in transferring Fukushima-derived radiocesium to a coniferous forest floor. Sci Total Environ 490:435–439. https://doi.org/10.1016/j.scitotenv.2014.05.034

The Ministry of Environment (2014) IAEA international follow-up mission final report. http://josen.env.go.jp/jishin/rmp/attach/iaea-finalrep140123_ja.pdf/

Yamaguchi N, Taniyama I, Kimura T, et al (2016) Contamination of agricultural products and soils with radiocesium derived from the accident at TEPCO Fukushima Daiichi Nuclear Power Station: monitoring, case studies and countermeasures. Soil Sci Plant Nutr:1–12. https://doi.org/10.1080/00380768.2016.1196119

Zaitsev AS, Gongalsky KB, Nakamori T, Kaneko N (2014) Ionizing radiation effects on soil biota: application of lessons learned from Chernobyl accident for radioecological monitoring. Pedobiologia (Jena) 57:5–14. https://doi.org/10.1016/j.pedobi.2013.09.005

Huang Y, Kaneko N, Nakamori T, Miura M, Tanaka Y, Nonaka M, Takenaka C (2016) Radiocesium immobilization to leaf litter by fungi during first-year decomposition in a deciduous forest in Fukushima. Journal of Environmental Radioactivity 152:28-34

Kaneko N, Huang Y, TNakamori T (2015) Decontamination of Radio Cs from Forest Soils Using Biodiversity and the Functioning of Soil Organisms. Journal of the Japanese Forest Society 97 (1):75-80

Chapter 11
Spiders as an Indicator of ^{137}Cs Dynamics in the Food Chains in Forests

Yoshiko Ayabe and Naoki Hijii

Abstract High concentrations of ^{137}Cs detected in the bodies of a web-building spider, *N. clavata*, collected at areas about 30 km apart from the FDNPP have indicated that contamination of the litter and upper-soil layers by radioactive fallout in the forest ecosystems in Fukushima has reached to the spider, a key predator working at a middle trophic level in food webs. In this chapter, we examined relationships between ^{137}Cs and alkali- and non-alkali metal elements measured for *N. clavata* and roles of alkali metals in the ^{137}Cs accumulations in the spider. Patterns and processes of the accumulation of ^{137}Cs closely associated with the alkali metals suggested the mechanisms of ^{137}Cs transfer and probable ^{137}Cs transfer pathways from soil to the web-building spider through the food chain. Understanding ^{137}Cs contamination in spiders would provide us crucial clues for clarifying the processes and mechanisms of ^{137}Cs transfer through food chains and for predicting the fate of ^{137}Cs in the forest ecosystems in Fukushima.

Keywords ^{137}Cs transfer · Food chains · *Nephila clavata* · Nuclear power plant accident · Radiocesium contamination · Web-building spider

11.1 Introduction

The accident at the Fukushima Daiichi Nuclear Power Plant (FDNPP) has wreaked large contaminated areas and hotspots of high radiation doses due to the radionuclide fallout in the forests in northeast Japan. Consequently, various organisms in forests and streams have been contaminated mainly with radiocesium (^{137}Cs) through the food chains (Murakami et al. 2014; Yoshimura and Akima 2014). The ecological processes involving a variety of organisms and the formation of complex food-web

Y. Ayabe (✉)
Institute for Environmental Sciences, Aomori, Japan
e-mail: ysk.ayabe@gmail.com

N. Hijii
Graduate School of Bioagricultural Sciences, Nagoya University, Nagoya, Aichi, Japan

C. Takenaka et al. (eds.), *Radiocesium Dynamics in a Japanese Forest Ecosystem*,
https://doi.org/10.1007/978-981-13-8606-0_11

structures in ecosystems work through two different lines of food chain (Polis and Strong 1996). In forest ecosystems, the grazing food chain starts with living plants, passing through herbivorous insects and other phytophages, and then reaches their predatory arthropods, such as spiders, and thereafter higher predators, such as sylvan birds. The detrital food chain starts with dead organic matter, such as plant litter suspended on the forest floor or in streams, and then energy flows via detritus feeders and microbes into their predators.

The half-life of ^{137}Cs as the major component of the radioactive fallout in the Fukushima accident is approximately 30.2 years; therefore, ^{137}Cs is expected to remain for a long time in the forest ecosystems in Fukushima. After the accident, adverse effects on animals inhabiting the contaminated area—for example, significantly higher rates of malformation and reduced survivorship in a gall-forming aphid and a lycaenid butterfly species—were observed (Akimoto 2014; Hiyama et al. 2012). Another observation suggests that the abundances of birds, butterflies, and cicadas decreased with increasing radiation dose in the contaminated area, although the abundance of spiders increased, probably due to the decrease in the abundance of birds, which are natural enemies of spiders (Møller et al. 2013). However, there is considerable debate concerning the effects of radiation on wildlife; relatively low rates of radiation, which were observed at most areas farther from seriously contaminated areas close to the FDNPP, are expected to be not so detrimental on animal populations (Beresford and Copplestone 2011). A reduced abundance of animals can depress ecosystem functions by deterring various biological interactions, such as pollination and predation pressure in regulating prey populations (Møller et al. 2012). The animals in the contaminated area not only could play an important role in sustaining ecosystem functions but also could be partly responsible for the ^{137}Cs circulation in the environment through the food chain. Insight into the patterns and processes of ^{137}Cs transfer from soil to plants and then to higher trophic levels can be provided by clarifying the mechanisms responsible for the ^{137}Cs contamination in animals; understanding these mechanisms is critically important in predicting the fate of ^{137}Cs in the forest ecosystems in Fukushima.

Spiders constitute over 50% of both number and biomass of carnivorous arthropods (Menhinick 1967), can function not only as the most effective predators of arthropods in both food chains (e.g., Moulder and Reichle 1972; Wise 1993) but also as major prey, along with insect caterpillars, for predators at higher trophic levels, such insectivorous bird species as Paridae (e.g., Kondo et al. 2017; Mizutani and Hijii 2002; Naef-Daenzer et al. 2000), and are thus key components of food webs in forest ecosystems. In the grazing food chain, adults of lepidopterans and sawflies, for example, whose larvae feed on living plants, could be captured by wondering or web-building spiders. In the detrital food chain, there are two pathways to reach spiders: one is that from aquatic insects, such as plecopterans, caddisflies, and dragonflies, emerging from the rivers and streams, where they feed on sedimentary plant litter or other aquatic insect larvae, while the other is that from detritivorous and fungivorous insects, such as chironomids, mycetophilids, and many other dipterans, emerging enormously from the litter in the tree canopy (Yoshida and Hijii 2005) and on the forest floor (Shimazaki and Miyashita 2005). Thus, understanding ^{137}Cs contamination in spiders would provide us a crucial clue for clarifying the processes and mechanisms of ^{137}Cs transfer through food chains in the forest ecosystems in Fukushima.

11.2 Radioactive Cs Contamination of Spiders

11.2.1 Spider

We focused on female adults of *Nephila clavata* L. Koch (Nephilidae: Arachnida). It is one of the largest web-building spiders in Japan and lives in an orb web set about 1–2 m above the ground. *Nephila clavata* is univoltine, overwinters as eggs, disperses as juveniles in April to May, and grows through seven or eight ecdyses to its largest size in September to October. The female spins an egg sack containing 500 to 1500 eggs after mating on the orb web in late August to October. It deposits the sack on trees or shrubs in October to November, and its life cycle ends by early winter. The body size is about 20 mm in female adults but no more than 10 mm in male adults. The size variability among mature female adults is large and depends on the amount of prey arthropods consumed (Miyashita 1992).

11.2.2 Sampling Sites

First sampling of spiders and measurement of air radiation dose rates were performed on 21 and 22 October 2012, about 1.5 years (19 months) after the FDNPP accident, at two sites in the Yamakiya district of Kawamata Town, and at another site in Koriyama City, in Fukushima Prefecture: (1) Site 1 (abbreviated as "PS" in Ayabe et al. (2014)) (ca. 1.6 ha), a streamside secondary forest dominated by broadleaved trees, such as *Quercus serrata*, *Q. crispula* (Fagaceae), and *Pterostyrax hispida* (Styracaceae), about 33 km northwest of the FDNPP, and faced southeastern on an abandoned pasture opening toward the FDNPP; (2) Site 2 (as "ES") (ca. 2.3 ha), a hillside secondary forest, consist of *Q. serrata*, *Ilex macropoda* (Aquifoliaceae), *Pourthiaea villosa* (Rosaceae), and some other shrub species, on the north side of the Yamakiya Elementary School about 37 km northwest of the FDNPP; and (3) Site 3 (as "KR"), an experimental pine forest in the Fukushima Prefectural Forestry Research Center in Koriyama City, about 62 km west of the FDNPP. As controls, spider individuals of the same species were collected on 30 October 2012 from Site CT, in a secondary forest on the main campus of Nagoya University, Nagoya, about 450 km in a straight line southwest of the FDNPP.

In 2013, the spider collection and soil sampling were made at Site 1 and Site 2 in late October, when female spiders maturated before oviposition (Ayabe et al. 2015a).

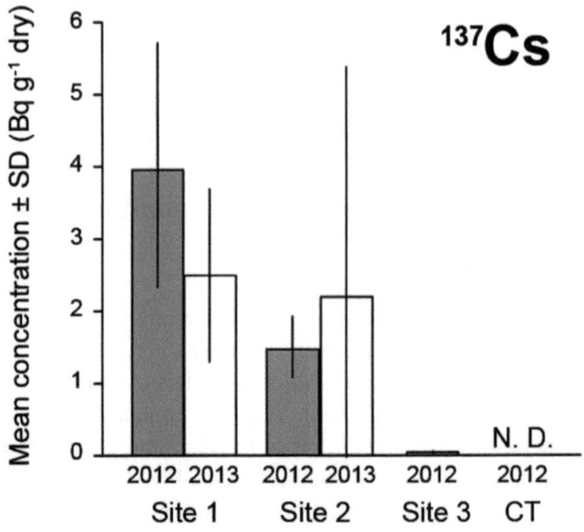

Fig. 11.1 Mean concentrations of radiocesium (^{137}Cs) in individuals of *N. clavata* collected at four different sites: Site 1, a riverside secondary forest 33 km northwest of the FDNPP; Site 2, a hillside secondary forest 37 km northwest of the FDNPP; Site 3, a pine forest 62 km west of the FDNPP; and CT, a secondary forest 450 km in a straight line southwest of the FDNPP as a control site. Vertical lines indicate the standard deviation. The data was derived from Ayabe et al. (2014 and 2015a)

11.2.3 Radiocesium Contamination of Spiders

In 2012, contamination with both radioactive ^{134}Cs (half-life 2.1 years) and ^{137}Cs was detected in all the spiders collected at both Site 1 and Site 2 (Fig. 11.1 represented for ^{137}Cs only). At Site 3, very low concentrations of ^{134}Cs and ^{137}Cs were detected only in a few individuals, and at Site CT, none was detected. The concentrations of ^{134}Cs and ^{137}Cs in spiders were highest at Site 1, followed by those at Site 2 and decreased with increasing of distance from the FDNPP (Ayabe et al. 2014).

The ratio of ^{134}Cs/^{137}Cs in 2012 corrected for radioactive decay to the date immediately after the FDNPP accident was 1.02 ± 0.19 on average (± SD) among samples of the three contaminated sites, which was equivalent to that of the fallout from the FDNPP (Haba et al. 2012). This clearly shown that the fallout from the FDNPP, not residual ^{137}Cs radioactivity derived from past nuclear weapon testing and the Chernobyl nuclear power plant accident (e.g., Fesenko et al. 2001), had contributed to the radiocesium contamination of spiders at these sites. *Nephila clavata* is a univoltine species, and all individuals were undoubtedly born after the accident at the FDNPP. Therefore, the contamination levels in spider individuals collected at the sites should reflect the contamination levels in their diet arthropods (most of which are also univoltine) at each site after the radioactive fallout.

Air radiation dose rates (μSv h^{-1}), which were correlated with concentration levels of radiocesium in soil (Nuclear Regulation Authority 2012), were measured with a scintillation survey 1 m aboveground level and were presented as an average of the measurements from 20 points (10 at upstream and 10 at downstream) at Site 1 and 10 points at Site 2 (Ayabe et al. 2014). The air radiation dose rate was highest at Site 1, followed by that at Site 2, and was much lower in Site 3 than these values,

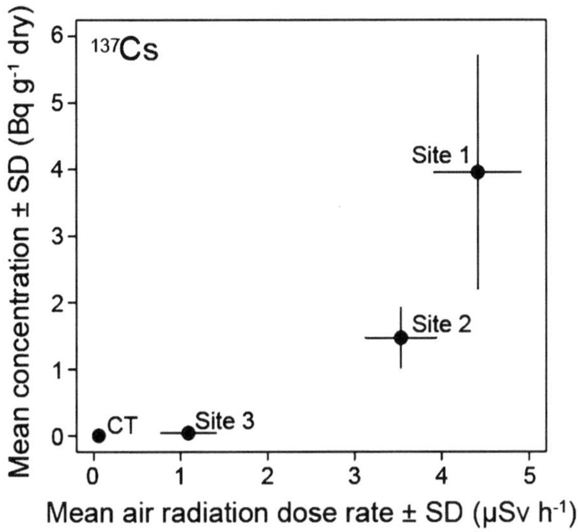

Fig. 11.2 Relationship between mean concentration of radiocesium (^{137}Cs) detected in *N. clavata* and mean air radiation dose rate at each site. Vertical and horizontal lines indicate the standard deviations of radiocesium concentration and air radiation dose rate, respectively. At Site 1, the air radiation dose rate was represented as the mean of 20 points; 10 points were used at the other sites. At site CT, radiocesium was not detected in spiders, and therefore, the mean value of radiocesium concentration was considered as 0 for practical purposes. (Modified from Ayabe et al. (2014))

and was close to zero at Site CT. At Site 2, ^{137}Cs deposition in soil (160 kBq m^{-2} dry) was less than half the amount at Site 1 (330 kBq m^{-2} dry; Ayabe et al. 2015a). At each site, the concentrations of ^{137}Cs in spiders tended to correlate positively with the air radiation dose rate (Fig. 11.2 presented for ^{137}Cs only) (Ayabe et al. 2014). Thus, the radioactive contamination levels in spiders tend to be associated with the radioactive contamination levels at the sites. At Site 1, two distant areas within the site had different air radiation dose rates (4.87 ± 0.19 at upstream, 3.96 ± 0.17 at downstream), suggesting that the radioactive contamination levels varied from place to place even within small areas of the same site.

The concentrations of ^{137}Cs in spider samples at Site 1 were 4.0 ± 1.8 (mean ± SD; Bq g^{-1} dry) in 2012 but decreased to 2.5 ± 1.2 in 2013, while at Site 2, 1.5 ± 0.45 in 2012 and 2.2 ± 3.2 in 2013 (Fig. 11.1; Ayabe et al. 2015a). The increase in mean concentration of ^{137}Cs at Site 2 in 2013 was due to a specific individual with an extremely higher concentration (ca. 9.0 Bq g^{-1} dry). It should be noted that this spider individual was collected at Site 2, which was less contaminated than Site 1, and even 2.5 years after the accident. In a species of wondering spiders, the biological half-life of ^{137}Cs is estimated at about 20 days and about 6 days for other herbivorous and predatory insects (Reichele 1967). If the value for *Nephila clavata* is similar to them, the specific individual could have foraged, by chance and soon before being collected, highly contaminated arthropods emerging at and flying from a radioactive hot spot elsewhere.

11.3 Mechanisms of ^{137}Cs Transfer to Spiders

11.3.1 Measurement of Element Contents in Spiders

Nephila clavata consumes arthropod prey caught on its orb web. The body size of mature female spiders depends on the amount of prey arthropods consumed (Miyashita 1991), but the ^{137}Cs concentration in *N. clavata* females was independent of the amount of prey consumed (Ayabe et al. 2014), suggesting that other mechanisms determine the level of ^{137}Cs accumulation in spider bodies. One factor contributing to ^{137}Cs contamination in spiders is uptake of potassium essential for plants and fungi at a lower trophic level. Because potassium and cesium have similar physiochemical traits, ^{137}Cs and the free hydrated cesium cation ($^{137}Cs^{+}$) can be absorbed from soil by plant roots and then translocated to the aboveground plant parts (Cid et al. 2013; Zhu and Smolders 2000). Uptake by living plants is the major pathway for transfer of ^{137}Cs from the soil to herbivores and then to arthropods and animals at higher trophic levels through the grazing food chain. ^{137}Cs concentration in animal tissues is suggested to be attributed to the level of Na- and K-ATPase activity (Kaikkonen et al. 2005), but no study has so far determined whether ^{137}Cs is assimilated into and eliminated from arthropods in a manner similar to that of other alkali metals.

Cs is an alkali metal and is expected to act similarly to Li, Na, K, and Rb. Animals including *N. clavata*, in particular, demand K and Na most among elements, and therefore, ^{137}Cs contamination should accompany the processes of accumulation of K and Na by *N. clavata*. If it is the case, ^{137}Cs concentrations in *N. clavata* would be correlated with the contents of K and Na. Stable cesium could be included in *N. clavata* prior to the accident and is expected to behave similarly to ^{137}Cs, and therefore, the patterns of ^{133}Cs accumulation could simulate the patterns of ^{137}Cs accumulation. Copper was highlighted as a key element among the non-alkali elements because (1) only Cu is known to have a physiological function (oxygen transport) in spiders (Terwilliger 1998) and (2) accumulation of Cu was correlated with the concentrations of many other non-alkali elements such as Ca and Zn (Ayabe et al. 2015b).

11.3.2 Relationship Between ^{137}Cs and Alkali- and Non-alkali Metal Elements in N. clavata

Nephila clavata contained K and Na in highest concentrations, followed by Mg, Ca, and Zn (Ayabe et al. 2015b). The contents of Rb, ^{133}Cs, and Li were substantially lower than that of K and Na. The relationship of ^{137}Cs accumulation with the other alkali metals was site specific. At the higher ^{137}Cs contamination site (Site 1), the accumulation of ^{137}Cs in *N. clavata* was associated with the contents of the alkali metals Na, K, and Rb, whereas the lower contamination site (Site 2) relative to Site

1, a significant correlation was detected only for the ^{137}Cs–Na combination (Fig. 11.3a). Among the alkali metals, the Na content only was correlated positively with ^{137}Cs accumulation at both sites. This suggests that ^{137}Cs accumulation in *N. clavata* could be more closely associated with the dynamics of Na, compared to the dynamics of K and Rb. Stable Cs also showed site-specific patterns; significant correlations between ^{133}Cs and K and Na or Rb were observed at Site 1, but not at Site 2, which was not the same as the pattern observed in ^{137}Cs (Fig. 11.3b). Moreover, ^{137}Cs–^{133}Cs correlation was significant at Site 1 only (Fig. 11.3c), and this suggests that ^{137}Cs and ^{133}Cs do not always behave similarly.

Cu concentration was significantly correlated with ^{137}Cs concentration at Site 1 but not at Site 2 (Fig. 11.3c). The correlation at Site 1, however, would not be attributed to a direct partial substitution of Cu with ^{137}Cs, because these two elements belong to different families. The Cu–^{137}Cs correlation may be related to the Cu–K relationship. Copper is essential for photosynthetic electron transport and accumulated in plants (Droppa et al. 1984), and Potassium is also essential for and absorbed by plants. Therefore, Cu and K are correlated at the plant level. Cu and K are both essential elements for arthropods as well; Cu is an essential component of hemocyanin that functions in the oxygen transport system in chelicerates (Burmester 2002; Terwilliger 1998) and can be involved in physical movement (muscle contraction), in which K is also critically important (Hoyle 1953; Katkowska 1995). In fact, the Cu–K correlation was observed in *N. clavata* collected at both sites (Ayabe et al. 2015b). K content was significantly correlated with ^{137}Cs concentration at Site 1 only, which might have resulted in an apparent correlation between Cu and ^{137}Cs at Site 1.

11.3.3 *Roles of Alkali Metals in the ^{137}Cs Accumulations in* N. clavata *and Probable ^{137}Cs Transfer Pathways from Soil to the Web-Building Spider*

The significant positive correlations between the ^{137}Cs concentration and the concentrations of other alkali metals in *N. clavata* at Site 1 indicated that ^{137}Cs behaved similarly to other alkali metals in the assimilation and elimination process in the bodies of *N. clavata*, which, in turn, determined the degree of ^{137}Cs contamination in *N. clavata*, at least at the higher contamination site (Site 1). However, correlations between ^{137}Cs and other alkali metals other than Na were not observed at the lower contaminated site (Site 2), and thus we concluded that ^{137}Cs contamination was attributed to the Na assimilation/elimination processes, but not always to the processes of other remaining alkali metals such as K and Rb. This difference in the degree of contribution to ^{137}Cs contamination by different alkali metals may be involved in the extent to which these elements are essential for plants and animals and their transfer through the food chains. The absorption of K is known to affect the accumulation of ^{137}Cs in plants because Cs, including ^{137}Cs, is absorbed through the

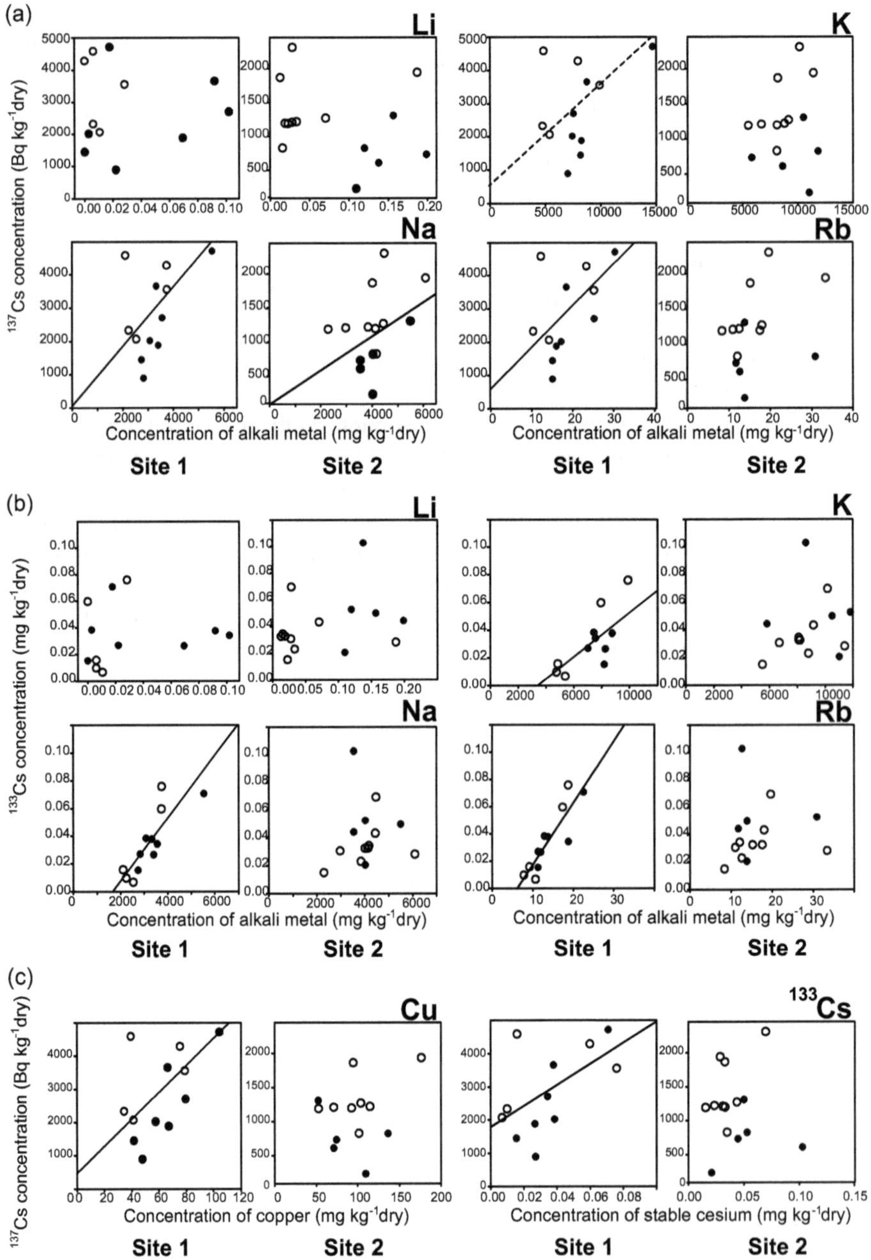

Fig. 11.3 Relationships between concentrations of the alkali metals (Li, Na, K, and Rb) and concentration of ^{137}Cs (**a**) and ^{133}Cs (**b**) and relationships between ^{137}Cs and Cu or ^{133}Cs (**c**) in *N. clavata* at Site 1 and Site 2. Open and filled circles indicate spider samples collected in 2012 and 2013, respectively. Solid lines show statistically significant relationships (GLMM; $p < 0.05$), and the dashed line shows a marginally significant relationship (GLMM; $p = 0.07$) (Modified from Ayabe et al. (2015b))

K transport system (Cid et al. 2013; Kabata-Pendias 2011; Zhu and Smolders 2000). The soil ^{137}Cs contamination level at Site 2 (160 ± 110 kBq m^{-2} dry in June 2013) was about half that at Site 1 (330 ± 120 kBq m^{-2} dry) (Ayabe et al. 2015a). Under a high concentration of K, plants become highly selective for K relative to ^{137}Cs (Zhu and Smolders 2000). Although K concentrations had not been evaluated at the two sites, we could expect a higher K/^{137}Cs concentration ratio at Site 2 unless extremely a lower K concentration was detected at Site 2. If this is true, plants might have been highly selective for K at Site 2. Such the K–^{137}Cs uptake pattern at the plant level might have caused nonsignificant relationships between K and ^{137}Cs in *N. clavata* at Site 2. Contrary to K, many higher plants have developed a high degree of selectivity for "not" uptaking Na, despite the chemical and physical similarities of K and Na (Subbarao et al. 2003). Even when Na is taken up by plants, it can be compartmentalized into the roots, with minimal transport to the shoots (Subbarao et al. 2003; Valdez-Aguilar and Reed 2008), i.e., a decreased concentration of Na in leaves and consequently limited transfer of Na up the food chain to animals. However, Na is essential for animals that demand Na as well as K. Therefore, to provide essential Na for higher animals, herbivorous insects must accumulate Na with a high efficiency from leaves with a decreased Na concentration (Seastedt and Crossley 1981). Crickets and spiders have high rates of Na assimilation (about 100%) but do not completely eliminate the element (e.g., Van Hook 1971). *Nephila clavata* in the present study accumulated a large amount of Na (Ayabe et al. 2015b). The essential requirement of Na in animals may have caused the significant correlation between Na and ^{137}Cs in *N. clavata* at both Site 1 and Site 2.

Uptake by plants is the entrance of the pathway of ^{137}Cs transfer from the soil to herbivores and then to arthropods and animals at higher trophic levels through the food chain. There was a large variation in ^{137}Cs contamination among plant species (Broadley and Willey 1997; Ertel and Ziegler 1991; Kuroda et al. 2013; Ramzaev et al. 2013; Yoshihara et al. 2013) and among seasons and plant organs even in a given plant species (von Fircks et al. 2002), probably due to different patterns of alkali metal accumulation. The ^{137}Cs depositions in soil were heterogeneous (Ayabe et al. 2015a, 2017), which may also contribute to differences in ^{137}Cs contamination among plant individuals. *Nephila clavata* feeds on a wide variety of arthropods, including herbivorous, detritivorous, and predatory arthropods. Many herbivorous species consume different plant species and possibly different plant individuals and parts within a plant species. These diverse feeding habits could cause a variation in ^{137}Cs contamination among herbivores, which would be passed on through the food chain to *N. clavata* at a higher trophic level and might result in a large variation in ^{137}Cs contamination among *N. clavata* individuals. Furthermore, *N. clavata* depends on both grazing and detritus food chains, and the relative dependency for these two food chains may also be involved in the site-specific ^{137}Cs contamination pattern and the among-individual variation in ^{137}Cs contamination. A better understanding of the transfer of ^{137}Cs through the food chain will require evaluation of the extents to which K and Na accumulation vary among plant species and tissues and also among various herbivorous arthropods with different feeding habits.

11.4 Spiders as an Indicator of ^{137}Cs Transfer in Forest Ecosystems

High concentrations of ^{137}Cs detected in the bodies of *N. clavata* have indicated that radioactive contamination of the litter and upper-soil layers in the forest ecosystem by radioactive fallout has reached to the spider, a key predator working at a middle trophic level, through the food chains. Wings and exoskeleton as residues of prey insects left on the spider web were mostly of aquatic insects, such as plecopterans and caddisflies, and detritivorous and fungivorous insects, such as chironomids, mycetophilids, and other dipterans. Most of them usually emerge from the leaf litter suspended in the tree canopy (Yoshida and Hijii 2005), on the forest floor (Shimazaki and Miyashita 2005; Hashimoto et al. 2012), and in streams (Ohte et al. 2013; Murakami et al. 2014). This evidence and high doses of ^{137}Cs deposition on the litter and upper-soil layers suggest that, in a few years, at least, after the accident, transfer of ^{137}Cs to *N. clavata* via the detrital food chain in which enormous numbers of detritus feeders emerging from contaminated leaf litter and fungi (Murakami et al. 2014) might have been dominant than that via the grazing food chain in which herbivorous insects feeding on plants that may uptake ^{137}Cs from deeper soil layers. In a coniferous forest near a nuclear fuel reprocessing plant in England, concentrations of radionuclides including ^{137}Cs were significantly higher in detritus feeders than in other feeding groups, and they varied seasonally depending on the dietary menu (Copplestone et al. 1999). In the case of the Chernobyl accident, radioactive contamination was clearly greater in detritus insects than in herbivorous and predatory insects in a grassland (Rudge et al. 1993), but the radioactive fallout had different effects on litter faunae, depending on the distance from the nuclear power plant and on the time since the accident (Krivolutzkii and Pokarzhevskii 1992).

The cause of the higher ^{137}Cs accumulation in spiders relative to other predators is associated also with the feeding manner; over 90% of ^{137}Cs absorbed by prey insects is distributed in their soft tissues, such as muscle and digestive tract that spiders ingest with digestive enzymes, while less than 1% is distributed in the exoskeleton that spiders leave intact (Cavalloro 1967; Copplestone et al. 1999). ^{137}Cs aggregated transfer coefficients from soil to plants may vary with the soil type (Calmon et al. 2009). Given different sites with the same soil type, soil-to-spider transfer coefficients are expected to be similar values as well. In this study, the soil type was the brown forest soil in both Site 1 and Site 2; the ^{137}Cs soil-to-spider transfer factors were similar between the sites (0.0066 for Site 1 and 0.0040 for Site 2; Ayabe et al. 2015a). These values were similar also to the ^{137}Cs concentration ratio of other spider species to contaminated soil in a coniferous forest near a nuclear fuel reprocessing plant in England (Toal et al. 2002). The dynamics of ^{137}Cs accumulation in the bodies of *N. clavata* feeding on detritus/plant feeders has evidenced that radioactive contamination transferred from litter/soil to arthropod predators through the detrital and grazing food chains, although the spider might have quantitatively much less impact on the circulation of ^{137}Cs within a forest ecosystem and/or on the diffusion out of the ecosystem than would do other larger predators with higher mobility.

Effects of radioactive contamination should be monitored successively on the population dynamics of organisms involved in the food web of a forest ecosystem. Since more than 20 years after the Chernobyl accident in 1986, declines of population levels associated with the high levels of radioactive contamination at their habitats have been apparent for spiders as well as insect and bird species (Møller and Mousseau 2006, 2009; Mousseau and Møller 2011), whereas spiders increased in number after the Fukushima accident due to the decrease in their enemy birds (Møller et al. 2013). Another study carried out in 1986 to 1988 reported that there was no significant effects of radioactive contamination on the population levels of soil animals in soil layers, in contrast to those in the litter layer (Krivolutzkii and Pokarzhevskii 1992). There has ever been much less information about impacts of radioactive contamination on the population levels of various organisms in forests contaminated by the Fukushima accident with a lower radioactive dose, as a whole, relative to the Chernobyl accident (Møller et al. 2013).

Our field survey revealed that detritus feeders emerging from the radioactive Cs pool could be preyed upon by spider communities, including *N. clavata*. Such spiders, in turn, as well as phytophagous insects, might then be foraged by avian predators at a higher trophic level, such as Paridae, and ^{137}Cs ingested by such prey arthropods are thus to be transferred to the higher trophic level (Fig. 11.4). This suggests that spiders are key predators connecting the grazing and detrital food chains in the food webs in forest ecosystems.

There are several other studies of radiocesium contamination in invertebrate species in Fukushima. *Nephila clavata* was more heavily contaminated with ^{134}Cs and ^{137}Cs than herbivorous insects, such as the rice grasshopper, *Oxya yezoensis*, and the Emma field cricket, *Teleogryllus emma* (Tanaka et al. 2016). In comparisons of ^{137}Cs contamination among different functional groups, such as herbivores, carnivores, omnivores, and detritivores, collected approximately 1.5–2.5 years after the accident, ^{137}Cs activity concentrations were low in herbivorous species, while high in detritivorous species that feed on plant litter and fungi of high ^{137}Cs concentration activity (Murakami et al. 2014; Ishii et al. 2017). A similar trend was also true of invertebrate species in other countries (Rudge et al.1993; Copplestone et al. 1999; Wood et al. 2009). Thus, the extent to which invertebrates are contaminated with ^{137}Cs is likely to be affected by what they consume as diets. Epigeic earthworms consume litter and soil and thus have shown relatively high values of transfer factors ranging from 0.21 to 0.35 (Hasegawa et al. 2013), compared to 0.006 in *N. clavata* (Ayabe et al. 2015a). These values, however, could not directly be comparable because the former was calculated using radiocesium concentrations in litter and soil dry mass (Bq g^{-1}) while the latter using radiocesium depositions in soil per unit area (Bq m^{-2}). A large fraction of ^{137}Cs contamination in epigeic earthworms was attributable to the habitat media (soil and litter) ingested in the intestine, and therefore gut clearance resulted in fast reduction of ^{137}Cs contamination; the biological half-life of ^{137}Cs in earthworms was expressed as dual exponential functions, where the half-life in a rapid loss due to gut clearance was estimated as

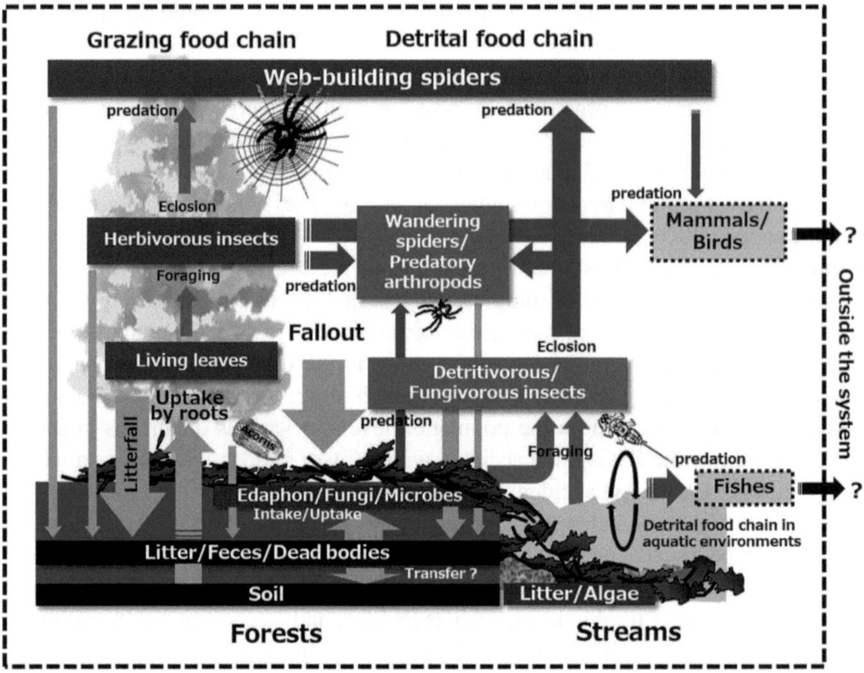

Fig. 11.4 Presumable processes on the transfer and circulation of radionuclides through the food chains in forest ecosystems. In the contaminated forest of Fukushima, ^{137}Cs taken by detritus feeders and their predatory arthropods were transferred to the web-building spider, *N. clavata*, through the detrital food chain, while ^{137}Cs taken up by living plants from the litter and soil could be transferred to the spider via herbivorous insects through the grazing food web. (Ayabe unpublished data)

0.1 day and that in a second slower loss due to physiological clearance as 27.4 days (Tanaka et al. 2018). These facts suggest that ^{137}Cs is unlikely to be highly bioaccumulated in the body wall muscle of earthworm (Tanaka et al. 2018). Since leaf litter and fungi have shown high levels of ^{137}Cs concentration activity, ^{137}Cs contamination levels would decrease with ascending the trophic level in detrital food chains from invertebrates to reptiles and fishes (Murakami et al. 2014). ^{137}Cs have moved from litter to soil layer 2–3 years after the accident, and a larger fraction of ^{137}Cs could be retained in the top soil layer, not in the litter layer (e.g., Ayabe et al. 2017; Fujii et al. 2014; Koarashi et al. 2016a, b; Nakanishi et al. 2014), and therefore, the amount of ^{137}Cs transfer through detrital food chains may have been reduced.

In future studies, we need to accumulate more data of the radiocesium concentrations of plant litter and detritus feeders in both the terrestrial and aquatic compartments (e.g., Avery 1996) and their transfer factors. Further information about the radiocesium concentrations in phytophagous insects feeding on live plants and seeds, the diet menus of these herbivores, and the relative proportions of arthropods preyed by predators such as spiders through between grazing and detrital food chains, by using stable isotopes (Haraguchi et al. 2013), for example, are required to evaluate the relative rates of radiocesium flow passing through diverse biological pathways in forest ecosystems. Forest ecosystems are more complex in species diversity and spatial structure than many other terrestrial ecosystems, and therefore, greater variability could be observed in radioactive transfer processes (e.g., Calmon et al. 2009). For the evaluation of effects of radioactive contamination on the ecosystem functions of forests, we need to examine the detrimental effects of radionuclide intake on the development and reproduction of food-web organisms (e.g., Mousseau and Møller 2011), which would in turn affect the population dynamics of the organisms and spatiotemporal patterns of the food-web structures.

The amount of ^{137}Cs transfer was suggested to differ between forests and streams, although the detrital-based food webs were similar between the subsystems (Sakai et al. 2016): ^{137}Cs concentration in stream litter was four times lower than in forest litter likely due to ^{137}Cs leaching from litter in streams, and consequently the ^{137}Cs concentration in the animal community differed between them. On the other hand, in a river in Fukushima, the total radioactive concentration in larvae of an aquatic insect (Stenopsychidae) feeding on fine particle organic matter was detected to reach >1000 Bq kg^{-1} dry in 2013 (Fujino et al. 2018).

Thus, we need long-term studies of different functional groups in both grazing and detrital-based food webs and in both terrestrial and aquatic subsystems in Fukushima forests to clarify the processes and mechanisms of radiocesium transfer through the food webs of forest ecosystems.

11.5 Materials and Methods

11.5.1 Evaluation of Radiocesium (^{137}Cs) Activity in Spider Samples

All spider samples were dried, and then the dry mass was individually weighed on a microbalance. Each dried body was put into a polyethylene tip with 5-mm-diameter zirconia beads and ground to powder with a bead-beater-type homogenizer. Concentration of radiocesium in the spider sample (Bq g^{-1} dry) was individually measured by a gamma spectrometer with a well-type germanium semiconductor detector (GWL-300-15-S, Ortec) (see Ayabe et al. 2014 for details).

11.5.2 Evaluation of Contents of the Alkali Metals in Spider Samples

Subsamples (ca. 0.1 g dry weight) of the spiders used for the ^{137}Cs measurement were digested with nitric acid solution (10 mL of concentrated nitric acid plus 5 mL of 5% nitric acid) at 160°C for 1 h and at 220°C 2 h until the samples were dissolved (Ayabe et al. 2015b). For each dissolved sample diluted to 50 mL with ultrapure water, the concentrations of the alkali metals (Li, Na, K, Rb, and ^{133}Cs) and Cu as a non-alkali metal were determined using the inductively coupled plasma mass spectrometry (iPAC Q, Thermo Scientific). The details of experimental procedures were described in Ayabe et al. (2015b).

Acknowledgment We thank Dr. H. Ozawa (Fukushima Prefectural Government) and Mr. T. Kanno (Kawamata, Fukushima Prefecture) for helpful support. We also thank Dr. Y. Ogata (Nagoya University) for helpful advices for our manuscript and Dr. R. Tomioka and Dr. Y. Sugiura (Japan Atomic Energy Agency) for the help with the elemental analyses. This study was supported by a Grant-in-Aid for Scientific Research from the Ministry of Education, Culture, Sports, Science, and Technology, Japan (no. 24110007) and in part by a Grant for Environmental Research Projects 2014 from the Sumitomo Foundation.

References

Akimoto S (2014) Morphological abnormalities in gall-forming aphids in a radiation-contaminated area near Fukushima Daiichi: selective impact of fallout? Ecol Evol 4:355–369. https://doi.org/10.1002/ece3.949

Avery SV (1996) Fate of caesium in the environment: distribution between the abiotic and biotic components of aquatic and terrestrial ecosystems. J Environ Radioact 30:139–171

Ayabe Y, Kanasashi T, Hijii N, Takenaka C (2014) Radiocesium contamination of the web spider *Nephila clavata* (Nephilidae: Arachnida) 1.5 years after the Fukushima Dai-ichi nuclear power plant accident. J Environ Radioact 127:105–110. https://doi.org/10.1016/j.jenvrad.2013.10.010

Ayabe Y, Kanasashi T, Hijii N, Takenaka C (2015a) Monitoring radiocesium contamination of the web spider *Nephila clavata* (Nephilidae: Arachnida) in Fukushima forests. J Jpn For Res 97:70–74. (in Japanese with English summary)

Ayabe Y, Kanasashi T, Hijii N, Takenaka C (2015b) Relationship between radiocesium contamination and the contents of various elements in the web spider *Nephila clavata* (Nephilidae: Arachnida). J Environ Radioact 150:228–235. https://doi.org/10.1016/j.jenvrad.2015.09.002

Ayabe Y, Hijii N, Takenaka C (2017) Effects of local-scale decontamination in a secondary forest contaminated after the Fukushima nuclear power plant accident. Environ Pollut 228:344–353. https://doi.org/10.1016/j.envpol.2017.05.041

Beresford NA, Copplestone D (2011) Effects of ionizing radiation on wildlife: what knowledge have we gained between the Chernobyl and Fukushima accidents? Integr Environ Assess Manag 7:371–373. https://doi.org/10.1002/ieam.238

Broadley MR, Willey NJ (1997) Differences in root uptake of radiocaesium by 30 plant taxa. Environ Pollut 97:11–15

Burmester T (2002) Origin and evolution of arthropod hemocyanins and related proteins. J Comp Physiol B Biochem Syst Environ Physiol. https://doi.org/10.1007/s00360-001-0247-7

Calmon P, Thiry Y, Zibold G, Rantavaara A, Fesenko SV (2009) Transfer parameter values in temperate forest ecosystems: a review. J Environ Radioact 100:757–766

Cavalloro R (1967) Data on the accumulation, distribution and elimination of Sr-85, I-131, and Cs-137 in various insect species. In: Aberg B, Hungate FP (eds) Radiological concentration processes. Pergamon Press, pp 601–608

Cid AS, Anjos RM, Zamboni CB, Velasco H, Macario K, Rizzotto M, Medeiros IMA, Juri Ayub J, Audicio P (2013) Temporal evolution of $^{137}Cs^{+}$, K^{+} and Na^{+} in fruits of South American tropical species. Sci Total Environ 444:115–120

Copplestone D, Jhonson MS, Jones SR, Toal ME, Jackson D (1999) Radionuclide behaviour and transport in a coniferous woodland ecosystem: vegetation, invertebrates and wood mice, *Apodemus sylvaticus*. Sci Total Environ 239:95–109

Droppa M, Terry N, Horvath G (1984) Effects of Cu deficiency on photosynthetic electron transport. Proc Natl Acad Sci U S A 81:2369–2373. https://doi.org/10.1073/pnas.81.8.2369

Ertel J, Ziegler H (1991) Cs-134/137 contamination and root uptake of different forest trees before and after the Chernobyl accident. Radiat Environ Biophys 30:147–157. https://doi.org/10.1007/BF01219349

Fesenko SV, Soukhova NV, Sanzharova NI, Avila R, Spiridonov SI, Klein D, Badot PM (2001) ^{137}Cs availability for soil to understory transfer in different types of forest ecosystems. Sci Total Environ 269:87–103

Fujii K, Ikeda S, Akama A, Komatsu M, Takahashi M, Kaneko S (2014) Vertical migration of radiocesium and clay mineral composition in five forest soils contaminated by the Fukushima nuclear accident. Soil Sci Plant Nutr:1–14. https://doi.org/10.1080/00380768.2014.926781

Fujino T, Kobori S, Nomoto T, Sakai M, Gomi T (2018) Radioactive cesium contamination and its biological half-life in larvae of *Stenopsyche marmorata* (Trichoptera: Stenopsychidae). Landsc Ecol Eng 14:37–43

Haba H, Kanaya J, Mukai H, Kambara T, Kase M (2012) One-year monitoring of airborne radionuclides in Wako, Japan, after the Fukushima Dai-ichi nuclear power plant accident in 2011. Geochem J 46:271–278

Haraguchi TF, Uchida M, Shibata Y, Tayasu I (2013) Contributions of detrital subsidies to aboveground spiders during secondary succession, revealed by radiocarbon and stable isotope signatures. Oecologia 171:935–944

Hasegawa M, Ito MT, Kaneko S, Kiyono Y, Ikeda S, Makino SI (2013) Radiocesium concentrations in epigeic earthworms at various distances from the Fukushima Nuclear Power Plnat 6 months after the 2011 accident. J Environ Radioact 126:8–13

Hashimoto S, Ugawa S, Nanko K, Shichi K (2012) The total amounts of radioactively contaminated materials in forests in Fukushima, Japan. Sci Rep 2:416. https://doi.org/10.1038/srep00416

Hiyama A, Nohara C, Kinjo S, Taira W, Gima S, Tanahara A, Otaki JM (2012) The biological impacts of the Fukushima nuclear accident on the pale grass blue butterfly. Sci Rep 2:570. https://doi.org/10.1038/srep00570

Hoyle G (1953) Potassium ions and insect nerve muscle. J Exp Biol 30:121–135

Ishii Y, Hayashi S, Takamura N (2017) Radiocesium transfer in forest insect communities after the Fukushima dai-ichi nuclear power plant accident. PLoS One 12(1):e0171133

Kabata-Pendias A (2011) Trace elements in soils and plants, 4th edn. CRC Press, Boca Raton

Kaikkonen M, de Gritz B, Eriksson L (2005) Short-term distribution of Cs in relation to Cr-EDTA after intravenous dose in goats. Acta Physiol Scand 183:321–332

Katkowska MJ (1995) The effect of external cesium ions on the muscle fibre resting potential in mealworm larva (*Tenebrio molitor* L.). J Comp Physiol A 177:519–526. https://doi.org/10.1007/BF00187487

Koarashi J, Atarashi-Andoh M, Matsunaga T, Sanada Y (2016a) Forest type effects on the retention of radiocesium in organic layers of forest ecosystems affected by the Fukushima nuclear accident. Sci Rep 6:1–11. https://doi.org/10.1038/srep38591

Koarashi J, Nishimura S, Nakanishi T, Atarashi-Andoh M, Takeuchi E, Muto K (2016b) Post-deposition early-phase migration and retention behavior of radiocesium in a litter–mineral soil system in a Japanese deciduous forest affected by the Fukushima nuclear accident. Chemosphere 165:335–341. https://doi.org/10.1016/j.chemosphere.2016.09.043

Kondo T, Mizutani M, Hijii N (2017) Small patches of broadleaf trees influence nest-site selection and reproductive performance of two tit species (Paridae) in a Japanese cedar plantation. J For Res 22:15–21

Krivolutzkii DA, Pokarzhevskii AD (1992) Effects of radioactive fallout on soil animal populations in the 30 km zone of the Chernobyl atomic power station. Sci Total Environ 112:69–77

Kuroda K, Kagawa A, Tonosaki M (2013) Radiocesium concentrations in the bark, sapwood and heartwood of three tree species collected at Fukushima forests half a year after the Fukushima Dai-ichi nuclear accident. J Environ Radioact 122:37–42. https://doi.org/10.1016/j.jenvrad.2013.02.019

Menhinick EF (1967) Structure, stability, and energy flow in plants and arthropods in a *Sericea lespedeza* stand. Ecol Monogr 37:255–272

Miyashita T (1991) Direct evidence of food limitation for growth rate and body size in the spider *Nephila clavata*. Acta Arachnol 40:17–21. https://doi.org/10.2476/asjaa.40.17

Miyashita T (1992) Variability in food consumption rate of natural populations in the spider, *Nephila clavata*. Res Popul Ecol 34:15–28

Mizutani M, Hijii N (2002) The effects of arthropod abundance and size on the nestling diet of two *Parus* species. Ornithol Sci 1:71–80. https://doi.org/10.2326/osj.1.71

Møller AP, Mousseau TA (2006) Biological consequences of Chernobyl: 20 years on. Tre Ecol Evol 21:200–207

Møller AP, Mousseau TA (2009) Reduced abundance of insects and spiders linked to radiation at Chernobyl 20 years after the accident. Bio Let Roy Soc 5:356–259

Møller AP, Barnier F, Mousseau TA (2012) Ecosystems effects 25 years after Chernobyl: pollinators, fruit set and recruitment. Oecologia 170:1155–1165. https://doi.org/10.1007/s00442-012-2374-0

Møller AP, Nishiumi I, Suzuki H, Ueda K, Mousseau TA (2013) Differences in effects of radiation on abundance of animals in Fukushima and Chernobyl. Ecol Indic 24:75–81. https://doi.org/10.1016/j.ecolind.2012.06.001

Moulder BC, Reichle DE (1972) Significance of spider predation in the energy dynamics of forest-floor arthropod communities. Ecol Monogr 42:473–498

Mousseau TA, Møller AP (2011) Landscape portrait: a look at the impacts of radioactive contaminants on Chernobyl's wildlife. Bull Atomic Sci 67:38–46

Murakami M, Ohte N, Suzuki T, Ishii N, Igarashi Y, Tanoi K (2014) Biological proliferation of cesium-137 through the detrital food chain in a forest ecosystem in Japan. Sci Rep 4:3599. https://doi.org/10.1038/srep03599

Naef-Daenzer L, Naef-Daenzer B, Nager RG (2000) Prey selection and foraging performance of breeding Great Tits *Parus major* in relation to food availability. J Avian Biol 31:206–214

Nakanishi T, Matsunaga T, Koarashi J, Atarashi-Andoh M (2014) 137Cs vertical migration in a deciduous forest soil following the Fukushima Dai-ichi Nuclear Power Plant accident. J Environ Radioact 128:9–14. https://doi.org/10.1016/j.jenvrad.2013.10.019

Nuclear Regulation Authority (2012) Summarized version of the "Results of the Research on Distribution of Radioactive Substances Discharged by the Accident at TEPCO's Fukushima Dai-ichi NPP". http://radioactivity.nsr.go.jp/en/contents/1000/294/view.html

Ohte N, Murakami M, Suzuki T, Iseda K, Tanoi K, Ishii N (2013) Diffusion and transportation dynamics of 137Cs deposited on the forested area in Fukushima after the Fukushima Daiichi Nuclear Power Plant Accident in March 2011. In: Nakanishi TM, Tanoi K (eds) Agricultural implications of the Fukushima nuclear accident. Springer, Singapore, pp 177–186

Polis GA, Strong DR (1996) Food web complexity and community dynamics. Am Nat 141:813–846

Ramzaev V, Barkovsky A, Goncharova Y, Gromov A, Kaduka M, Romanovich I (2013) Radiocesium fallout in the grasslands on Sakhalin, Kunashir and Shikotan Islands due to Fukushima accident: the radioactive contamination of soil and plants in 2011. J Environ Radioact 118:128–142. https://doi.org/10.1016/j.jenvrad.2012.12.006

Reichele DE (1967) Relation of body size to food intake, oxygen consumption, and trace element metabolism in forest floor arthropods. Ecology 49:538–542

Rudge SA, Johnson MS, Leah RT, Jones SR (1993) Biological transport of radiocesium in a semi-natural grassland ecosystem. 1. Soils, vegetation and invertebrates. J Environ Radioact 19:173–198

Sakai M, Gomi T, Negishi JN, Iwamoto A, Okada K (2016) Different cesium137 transfers to forest and stream ecosystems. Environ Pollut 209:46–52
Seastedt TR, Crossley DA (1981) Sodium dynamics in forest ecosystems and the animal starvation hypothesis. Am Nat 117:1029–1034
Shimazaki A, Miyashita T (2005) Variable dependence on detrital and grazing food webs by generalist predators: aerial insects and web spiders. Ecography 28:485–494
Subbarao GV, Ito O, Berry WL, Wheeler RM (2003) Sodium – a functional plant nutrient. CRC Crit Rev Plant Sci 22:391–416. https://doi.org/10.1080/748638747
Tanaka S, Hatakeyama K, Takahashi S, Adati T (2016) Radioactive contamination of arthropods from different trophic levels in hilly and mountainous areas after the Fukushima Daiichi nuclear power plant accident. J Environ Radioact 164:104–112
Tanaka S, Adati T, Takahashi T, Fujiwara K, Takahashi S (2018) Concentrations and biological half-life of radioactive cesium in epigeic earthworms after the Fukushima Dai-ichi Nuclear Power Plant accident. J Environ Radioact 192:227–232
Terwilliger N (1998) Functional adaptations of oxygen-transport proteins. J Exp Biol 201:1085–1098
Toal ME, Copplestone D, Johnson MS, Jackson D, Jones SR (2002) Quantifying ^{137}Cs aggregated transfer coefficients in a semi-natural woodland ecosystem adjacent to a nuclear reprocessing facility. J Environ Radioact 63:85–103
Valdez-Aguilar LA, Reed DW (2008) Influence of potassium substitution by rubidium and sodium on growth, ion accumulation, and ion partitioning in bean under high alkalinity. J Plant Nutr 31:867–883. https://doi.org/10.1080/01904160802043213
Van Hook RJ (1971) Energy and nutrient dynamics of spider and orthopteran populations in a grassland ecosystem. Ecol Monogr 41:1–26
Von Fircks Y, Rosen K, Sennerby-Forsse L (2002) Uptake and distribution of Cs-137 and Sr-90 in *Salix viminalis* plants. J Environ Radioact 63:1–14. https://doi.org/10.1016/S0265-931X(01)00131-X
Wise DH (1993) Spiders in ecological webs. Cambridge University Press, Cambridge, UK. https://doi.org/10.1017/CBO9780511623431
Wood M, Leah R, Jones SR, Copplestone D (2009) Radionuclide transfer to invertebrates and small mammals in a coastal sand dune ecosystem. Sci Total Environ 407:4062–4074
Yoshida T, Hijii N (2005) The composition and abundance of microarthropod communities on arboreal litter in the canopy of *Cryptomeria japonica* trees. J For Res 10:35–42
Yoshihara T, Matsumura H, Hashida S, Nagaoka T (2013) Radiocesium contaminations of 20 wood species and the corresponding gamma-ray dose rates around the canopies at 5 months after the Fukushima nuclear power plant accident. J Environ Radioact 115:60–68. https://doi.org/10.1016/j.jenvrad.2012.07.002
Yoshimura M, Akima A (2014) Radioactive contamination of aquatic insects in a stream impacted by the Fukushima nuclear power plant accident. Hydrobiologia 722:19–30
Zhu YG, Smolders E (2000) Plant uptake of radiocaesium: a review of mechanisms, regulation and application. J Exp Bot 51:1635–1645. https://doi.org/10.1093/jexbot/51.351.1635

[illegible]

[illegible]

[illegible]

Sutton [illegible] (19[illegible]) [illegible] plant material [illegible]

[illegible]

[illegible] Nuclear Power Plant accident [illegible]

[illegible] (1982) [illegible] J Exp Biol [illegible] 1098 [illegible]

[illegible] M, Copplestone D, [illegible] MS, Jackson D, Jones SR (2003) Quantifying ^{137}Cs aggregated transfer coefficients in a semi-natural woodland ecosystem adjacent to a nuclear reprocessing facility. [illegible]

[illegible] (1995) [illegible] by earthworm and sediment [illegible]

[illegible]

[illegible]

[illegible] Total Environ [illegible] 2074

[illegible]

[illegible]

[illegible]

[illegible]

Chapter 12
Radioactive Cesium Contamination of Sika Deer in Oku-Nikko Region of Tochigi Prefecture in Central Japan

Masaaki Koganezawa, Kei Okuda, and Emiko Fukui

Abstract ^{137}Cs concentrations in the muscles and the rumen and rectal contents of sika deer and in dwarf bamboo and other plants as the main winter food source of the sika deer were investigated in the Oku-Nikko, Tochigi Prefecture, Japan. The ^{137}Cs concentration of the 5-year residual winter leaves of dwarf bamboo from 2012 decreased over the years, but the T_{ag} of dwarf bamboo from the soil to dwarf bamboo did not change. The average ^{137}Cs concentrations of the muscles of the deer decreased from the winter of 2012 to that of 2016, and those of the rumen contents and the rectal contents collected from 2012 to 2014 have decreased over time. The ^{137}Cs concentration of the rectal contents was 2.4 times that of the rumen contents and 3 times the muscle, suggesting that sika deer may excrete most of ^{137}Cs taken from foods while concentrating it in the digestive tract. The T_{ag} has been decreasing with year: from the dwarf bamboo to the deer muscles, 21.5–31.3 m^2 kg^{-1} dwt, and from the soil to the deer muscles, 0.02–0.01 m^2 kg^{-1} dwt.

Keywords Aggregated transfer coefficient (T_{ag}) · Dwarf bamboo · Muscle · ^{137}Cs transfer · Rectal contents · Rumen contents · Sika deer

12.1 Introduction

Patterns and processes of the transfer of ^{137}Cs from the soil to upper trophic levels through food webs are one of the key aspects that should be addressed to clarify the radionuclide dynamics in forest ecosystems. We sought to obtain basic quantitative data on the accumulation dynamics of the radioactive material in the components of

M. Koganezawa (✉)
Center for Weed and Wildlife Management, Utsunomiya University, Utsunomiya, Japan
e-mail: masaakik@cc.utsunomiya-u.ac.jp

K. Okuda
Faculty of Human Environmental Studies, Hiroshima Shudo University, Hiroshima, Japan

E. Fukui
School of Agriculture, Utsunomiya University, Utsunomiya, Japan

C. Takenaka et al. (eds.), *Radiocesium Dynamics in a Japanese Forest Ecosystem*,
https://doi.org/10.1007/978-981-13-8606-0_12

a soil-food (plants) and plants-sika deer (*Cervus nippon*) in a forest ecosystem. Furthermore, we attempted to elucidate the mechanism of contamination. We collected the leaves of the dwarf bamboo (*Sasa nipponica*) and other plants (*Quercus crispula* and/or *Larix kaempferi*), which are the main winter food source of the sika deer (Maruyama 1981) which were extracted for investigation. Furthermore, sika deer were captured from the Oku-Nikko, Tochigi Prefecture, Japan, and we collected and extracted the muscles of the biceps femoris, as well as the rumen and rectal content, and measured the ^{137}Cs concentration of each sample.

12.2 ^{137}Cs Transfer from Soil to Deer Via Dwarf Bamboo

The early stage ^{137}Cs deposition flux measured in the ground of the area investigated was an average of 8188 Bq m^{-2} in July 2011 (Kato personal communication). The biomass of leaf parts of the dwarf bamboo was 62.89 ± 32.00 g dwt m^{-2} (mean ± SD, $n = 40$) in November 2013. The ^{137}Cs concentration of the 5-year residual winter leaves of dwarf bamboo from 2012 decreased over the years from 101.3 ± 62.2 to 53.8 ± 45.0 Bq kg^{-1} dwt (mean ± SD). Moreover, no change was observed in the yearly aggregated transfer coefficient of dwarf bamboo from the soil to dwarf bamboo (T_{ag} = activity concentration in dwarf bamboo [Bq kg^{-1} dwt]/ amount in soil [Bq m^{-2}]), for 5 years at 0.01 m^2 kg^{-1} (Table 12.1). The average ^{137}Cs concentration in the muscles of the deer decreased from 155.2 ± 36.7 to 93.8 ± 56.5 Bq kg^{-1} dwt (mean ± SD) from the winter of 2012 to that of 2016.

Moreover, the T_{ag} has been decreasing with year: from the dwarf bamboo to the deer muscles, 21.5–31.3 m^2 kg^{-1} dwt, and from the soil to the deer muscles, 0.02–0.01 m^2 kg^{-1} dwt (Table 12.2). Furthermore, we determined the ^{137}Cs concentration and the T_{ag} from the ground to plant parts of oak, larch, and elm collected between 2012 and 2014 (Table 12.3). The results showed that the ^{137}Cs concentration and the T_{ag} of the bark of oak were highest, and those of the barks of larch and elm were higher than those of the dwarf bamboo. This suggests that the T_{ag} from the

Table 12.1 ^{137}Cs concentration of the winter leaf of *Sasa nipponica* and the transfer coefficient (Tag) in Oku-Nikko, Tochigi Prefecture, Japan, between 2012 and 2016

			Leaf of Sasa		Deposition flux	Transfer coefficient (Tag)	Leaf of Sasa per unit area
Year	Month	*n*	Bq kg^{-1} dwt	SD	Bq m^{-2}	m^2 kg^{-1} dwt	Bq m^{-2}
2012	Nov	9	101.3	62.2	8000	0.012	5.6
2013	Nov	10	101.2	66.0	7820	0.013	5.6
2014	Nov	10	88.0	23.8	7640	0.012	4.8
2015	Nov	8	107.1	76.4	7460	0.014	5.9
2016	Dec	7	53.8	45.0	7296	0.007	3.0

Table 12.2 ^{137}Cs concentration in the muscles of a deer and the transfer coefficient (Tag) in Oku-Nikko, Tochigi Prefecture, Japan, between 2012 and 2016

			Muscles of a deer		Sasa deer	Soil deer
Year	Month	*n*	Bq kg^{-1} dwt	SD	m^2 kg^{-1} dwt	m^2 kg^{-1} dwt
2012	Feb-Mar	17	155.2	36.7	–	0.02
2013	Feb-Mar	26	122.8	38.7	21.5	0.02
2014	Feb-Mar	26	132.6	35.1	27.6	0.02
2015	Apr	16	106.1	43.9	18.0	0.01
2016	Apr	9	93.8	56.5	31.3	0.01

Table 12.3 ^{137}Cs concentration of another food plants and a winter leaf of *Sasa nipponica* and the transfer coefficient (Tag) in Oku-Nikko, Tochigi Prefecture, Japan, between 2012 and 2014

Species/part/ (*n*)	Collected year	^{137}Cs [Bq kg^{-1} dwt]	The transfer coefficient (Tag) [m^2 kg^{-1} dwt]
Quercus crispula			
Fallen leaf (2)	Apr 2012	119.7	0.015
		140.1	0.018
Leaf (1)	Sep 2013	52.0	0.007
Nut (1)	Sep 2013	42.0	0.005
Bark (3)	May 2014	897.1	0.117
		1406.5	0.184
		2691.9	0.352
Larix kaempferi			
Young twig (2)	Apr 2012	37.1	0.005
		55.4	0.007
Bark (2)	Sep 2013	209.0	0.027
		244.0	0.031
Ulmus davidiana			
Young twig (2)	Apr 2012	245.0	0.031
		256.3	0.032
Bark (1)	Jun 2014	207.0	0.027
Sasa nipponica			
Leaf (9)	Nov 2012	101.3 ± 62.2 (SD)	0.013

soil and feed is higher in winter with heavy snow, as deer eat bark more in winter when there is high snow accumulation.

The ^{137}Cs concentration of the rumen contents collected from 2012 to 2014 has decreased from 258.8 ± 147.3 to 117.9 ± 86.5 Bq kg^{-1} dwt (mean ± SD) over time. On the other hand, the highest T_{ag} from rumen contents to the deer muscle was observed in 2014 (Table 12.4). In addition, it has been suggested that the T_{ag} possibly increased by feeding of many barks in consideration of the larger snow amount in 2014 (data provided by Japan Meteorological Agency) compared with other years. Furthermore, the ^{137}Cs concentration of the rectal contents collected from 2012 to 2014 has decreased from 489.0 ± 132.3 to 377.1 ± 141.5 Bq kg^{-1} dwt (mean ± SD) over time. Moreover, the ^{137}Cs concentration of the rectal content was

Table 12.4 ^{137}Cs concentration of rumen content and the transfer rate in Oku-Nikko, Tochigi Prefecture, Japan, between 2012 and 2014

Year	*n*	Rumen content [Bq kg^{-1} dwt]	SD	Muscles of a deer [Bq kg^{-1} dwt]	Transfer rate
2012	30	258.8	147.3	155.2	0.60
2013	46	175.5	96.9	122.8	0.70
2014	25	117.9	86.5	132.6	1.12

Table 12.5 ^{137}Cs concentration of rectal content in Oku-Nikko, Tochigi Prefecture, Japan, between 2012 and 2014

Year	*n*	Rectal content [Bq kg^{-1} dwt]	SD	Muscles of a deer [Bq kg^{-1} dwt]	Transfer rate
2012	40	489.0	132.3	155.2	0.32
2013	46	367.0	131.9	122.8	0.33
2014	27	377.1	141.5	132.6	0.35

2.4 times that of the rumen content and 3 times the muscle (Table 12.5). It has been suggested from the result that deer may excrete most of ^{137}Cs taken from foods while concentrating it in the digestive tract.

12.3 Changes in the ^{137}Cs Levels in Soil Depending on the Density of Deer Population

In order to examine reduction of ^{137}Cs from deer population to soil in the region in question, density of deer population was estimated from number of exterminated deer by the culling and results of a habitat density survey based on quadrat method conducted after the culling. The density of the deer population in the region during the winter of 2014 decreased based on the results of the culling that were performed in February, and the density was investigated in March 2014. Forty-seven deer were caught during the extermination performed on February 2015; the capture density was 12.4 deer km^{-2}. Moreover, 36 animals were found to inhabit the area based on the density survey conducted on March 2015, and the habitat density was 21.8 deer km^{-2}. These findings indicate that the habitat density before the culling was 34.2 deer km^{-2}, and the number of inhabitants was presumed to be 1269 animals.

In analyzing the soil, we must assume that deer carcasses and feces were present. The reduction in accumulation level because of the presence of deer carcasses was presumed to be 1183 Bq/animal based on the average weight (37.1 kg) and an average ^{137}Cs concentration of 31.9 Bq kg^{-1} fwt, fresh weight), in the muscles. On the other hand, because of the seasonal migration of the deer in this region, the

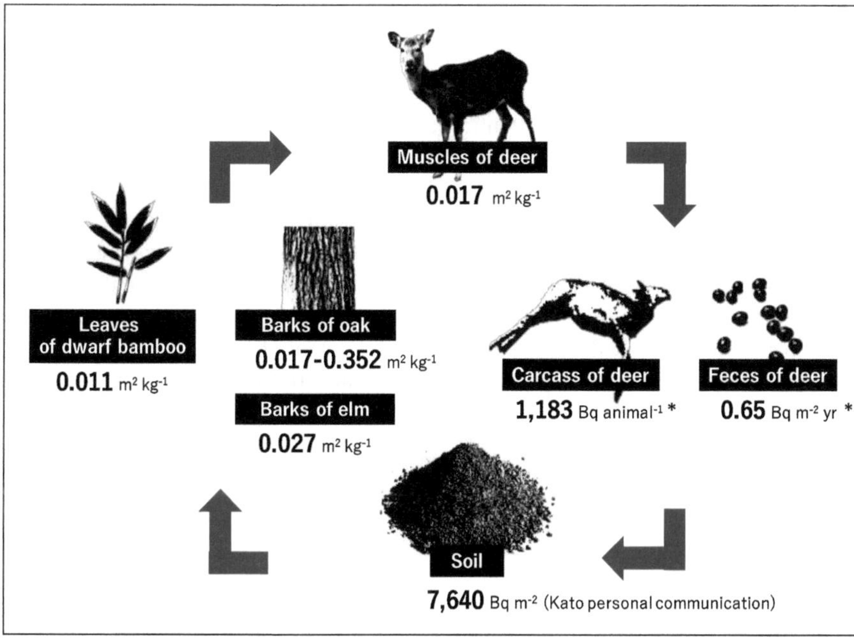

* The amount of radioactive cesium was calculated based on the density of deer in Oku-Nikko.

Fig. 12.1 T_{ag} of radioactive cesium concentration to deer's main foods (plants) from the soil and to deer muscle from the plants in Oku-Nikko, Tochigi Prefecture, central Japan, and the reduction of radioactive cesium from deer population to the soil in this area

density of the region in summer differs from that during the hibernation term. We determined the amount of contamination deposited in the ground through the deer feces based on the investigated density of 34.2 deer km^{-2} and 7.9 deer km^{-2} from the discharge amount (250 g dwt·day^{-1}·deer^{-1} [Takatsuki et al. 1981]). The total amount of excrement per unit area during the hibernation term was 1291 kg km^{-2}, and the amount of ^{137}Cs reduced by the excrement was 0.49 Bq m^{-2}. Moreover, the total amount of excrement during the settlement of individual deer in the summer was 422.65 kg km^{-2}, and the reduction in the amount of ^{137}Cs from the excrement was 0.16 Bq m^{-2}, and this value was 0.65 Bq m^{-2} by the following 1 year anniversary (Fig. 12.1).

After the accident at FNPP, many studies have been made in Japan on direct impacts of radiation on wild mammals such as impacts on health of Japanese macaque (*Macaca fuscata*) by low-dose exposure (e.g. Ochiai et al. 2014) as well as on indirect impacts on them associated with nuclear accident such as impacts on wild mammal population dynamics associated with people's evacuation (e.g. Fukasawa et al. 2016) in addition to this study on dynamics of transfer of cesium to wild mammals (e.g. Tagami et al. 2016).

12.4 Study Site

This study was conducted in the Oku-Nikko (elevation: 1300–1500 m, 36°45′N, 139°30′E) of Tochigi Prefecture, central Japan. In 2011, radioactive cesium (^{134}Cs + ^{137}Cs) fallout after the Fukushima nuclear accident reached 10,000–30,000 Bq m^{-2} in this area, located 160 km southwest of the FNPP. The study area was a gently sloping forest located between Mt. Nantai and Lake Chuzenji (elevation: 1269 m). The vegetation was a mixed broadleaf-conifer forest dominated by Japanese beech (*Fagus crenata*), oak (*Quercus crispula*), Nikko fir (*Abies homolepis*), and Japanese cherry birch (*Betula grossa*). Most of the understory vegetation was composed of dwarf bamboo, which had a high crude protein content (12.4%) even in the winter because it was an evergreen perennial (Seto et al. 2015). The plant had reduced height and biomass because of overgrazing (Seto et al. 2015).

References

Fukasawa K, Mishima Y, Yoshioka A, Kumada N, Totsu K, Osawa T (2016) Mammal assemblages recorded by camera traps inside and outside the evacuation zone of the Fukushima Daiichi nuclear power plant accident. Ecol Res 31:493. (data paper)

Maruyama N (1981) A study of the seasonal movements and aggregation patterns of Sika deer. Bull Fac Agric Tokyo Univ. Agric. Technol 23, 85 pp

Ochiai K, Hayama S, Nakiri S, Nakanishi S, Ishii N, Uno T, Kato T, Konno F, Kawamoto Y, Tsuchida OT (2014) Low blood cell counts in wild Japanese monkeys after the Fukushima Daiichi nuclear disaster. Sci Rep 4:5793

Seto T, Matsuda N, Okahisa Y, Kaji K (2015) Effects of population density and snow depth on the winter diet composition of sika deer. Wildlife Manag 79:243–253

Tagami K, Howard BJ, Uchida S (2016) The time-dependent transfer factor of radiocesium from soil to game animals in Japan after the Fukushima Daiichi nuclear accident. Environ Sci Technol 50:9424–9431

Takatsuki S, Koki K, Suzuki K (1981) Defecation rates of sika deer and Japanese serow. Jpn J Ecol 31:435–439

Part IV
Radiocesium Dynamics and Its Perspective in Forests

Chapter 13
Modeling Radiocesium Dynamics in a Contaminated Forest in Japan

Tsutomu Kanasashi

Abstract For the management of radionuclide contamination in a forest, a dynamic model for ^{137}Cs was developed applying previous study models and using explicit measured data in Fukushima area. The transfer of ^{137}Cs among organic layer, surface soil, and parts of trees was estimated in deciduous broad-leaved (konara; *Quercus serrata*) forest and evergreen coniferous needle-leaved Japanese cedar (sugi; *Cryptomeria japonica*) forest. Barks in both of the forest and surface of leaves were utilized as one of the compartments in this model because of the surface deposition of ^{137}Cs and the ^{137}Cs infiltration from the surfaces. Monte Carlo method was applied to optimize the best set of parameter values for mathematical equations, which control ^{137}Cs transfer between the compartments. The model estimates well fit to the observed data at konara and sugi stands in one place, but did not fit to that of both stands in another place because of the difference in concentration in wood part of both tree species.

Keywords Dynamic model · Konara oak (*Q. serrata*) · Sugi (*C. japonica*) · Monte Carlo method

13.1 Introduction

Radiocesium contamination in forest ecosystems will continue because of longer half-time of ^{137}Cs (30.2 years). Five years after the Fukushima Daiichi Nuclear Power Plant (FDNPP) accident, most of the ^{137}Cs was accumulated in soil (Imamura et al. 2017). Removing the soil could be the most effective decontamination, but could be impossible to remove that from the huge forest area, except for restricting to small area under any rule. Therefore, monitoring and predicting the contamination is

T. Kanasashi (✉)
Center for Research in Isotopes and Environmental Dynamics, University of Tsukuba, Tsukuba, Ibaragi, Japan
e-mail: kanasashi.tsutomu.gm@u.tsukuba.ac.jp

C. Takenaka et al. (eds.), *Radiocesium Dynamics in a Japanese Forest Ecosystem*,
https://doi.org/10.1007/978-981-13-8606-0_13

needed for the knowledge of future distribution of ^{137}Cs inventory in forest and management of the contamination.

In the case of Chernobyl Nuclear Power Plant (CNPP) accident, many models were developed to predict ^{137}Cs distribution in forest ecosystems (IAEA 2002). Some model predictions for ^{137}Cs dynamics in the forests have been developed after the FDNPP accident (Hashimoto et al. 2013; Nishina and Hayashi 2015). Hashimoto et al. (2013) applied the RIFE1 (Radionuclides in Forest Ecosystems) model, developed after the Chernobyl accident, to track radionuclide dynamics in forest ecosystems (IAEA 2013), and Nishina and Hayashi (2015) developed FoRothCs model, which is based on the Rothamsted Carbon model and the diameter distribution prediction system (DDPS). These models well fit the observed data in each subject forest site in eastern Japan, but did not validate their model adaptivities to other forest stands.

In this chapter, a ^{137}Cs dynamic model was developed to predict the temporal variation of ^{137}Cs inventory in forest ecosystems of deciduous broad-leaved trees dominated by konara (*Quercus serrata*) and evergreen coniferous trees of sugi (*Cryptomeria japonica*) in Fukushima. A simple dynamic model was consulted for prediction. The organic layer, surface soil, and some parts of the dominant trees were targeted to predict the ^{137}Cs distribution, excluding shrubs and game animals due to their small biomass in the forest ecosystems (Fig. 13.1 and Table 13.1). A method to develop the model is mentioned in Sect. 13.5. In this chapter, words to represent names of the compartment are shown using capital letters at their first alphabets in any time.

13.2 Estimate of ^{137}Cs Dynamic Model in Konara and Sugi

In both of the models for deciduous broad-leaved forest (konara model) and evergreen coniferous trees of sugi (sugi model), the estimates of ^{137}Cs fit the measured data for both the konara forest in Ohtama and the sugi forest in Kawauchi, Fukushima (Table 13.2), under the optimized parameter values tuned by the Monte Carlo method. The konara model well fits the measured data of Imamura et al. (2017). For both of the konara and sugi models, the temporal changes of ^{137}Cs estimates had been drastic for the initial 5 years after the FDNPP accident (Fig. 13.2), and then, little drastic variations were estimated from then, except for A0 compartments, which drastically decreased until approximately 10 years for the konara model and approximately 7 years for the sugi model. While, Wood was estimated to have smaller temporal change than the other compartments and kept a small amount of ^{137}Cs radioactivity during 30 years. In the konara model, A0 and Bark reached a similar amount of ^{137}Cs radioactivity (kBq m^{-2}), while, in the sugi model, A0 and Wood showed a similar one.

For the konara model, most of the component in A0 was composed of leaf litter without the initial deposition of ^{137}Cs after the drastic variations ceased, but some of the ^{137}Cs in A0 was not transferred to Soil in a season due to slow leaf

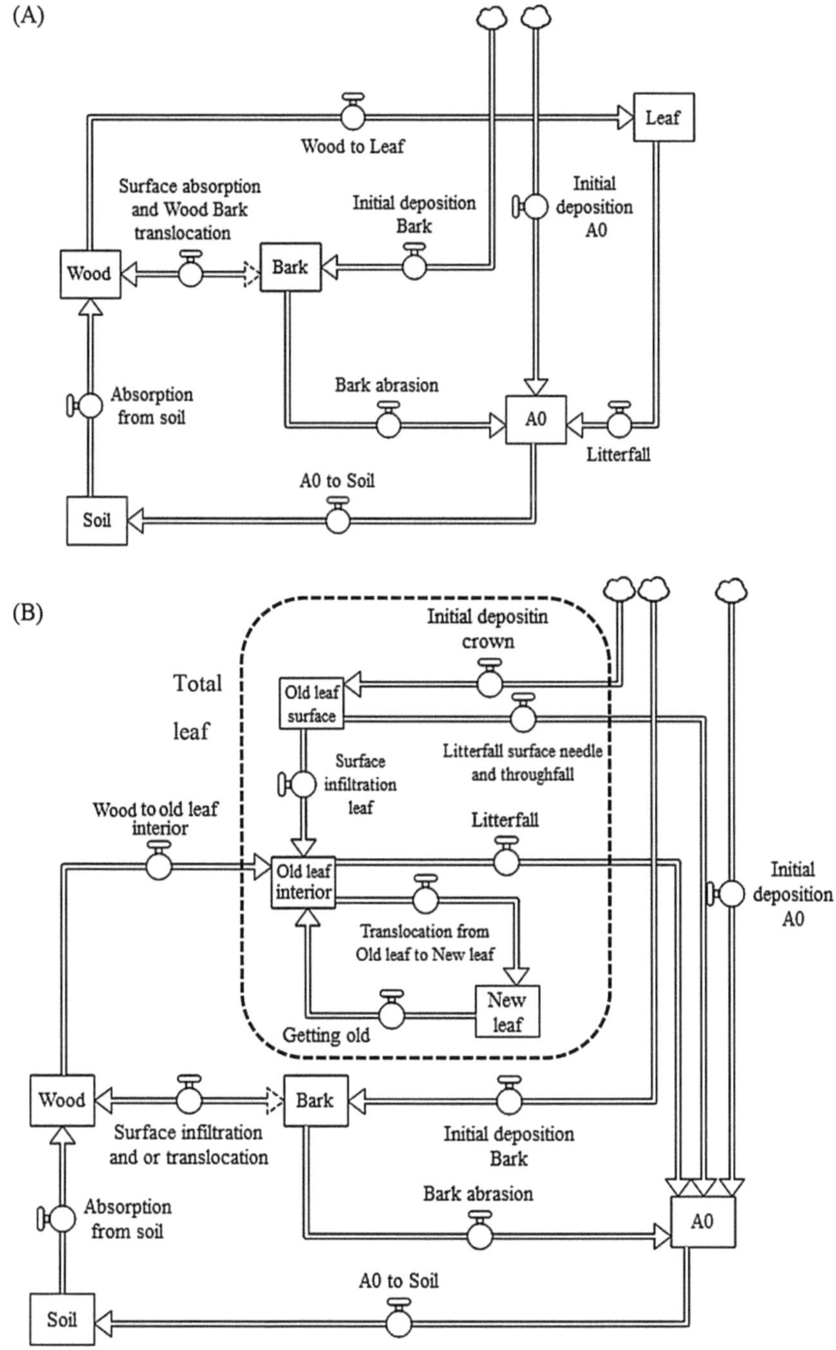

Fig. 13.1 Conceptual description of the dynamic model for (**a**) konara oak forest and (**b**) sugi cedar forest. Rectangles and unfilled arrows indicate model compartments and direction of ^{137}Cs flow controlled by the model equations symbolized by valves added to the arrows, respectively. The figures were created using STELLA Professional software (ver. 1.0.3)

Table 13.1 Compartments and related parameters used in the models of (a) konara and (b) sugi

Compartment	Description	Inflow/ outflow	Related parameter for translocation
(a) konara			
A0	Organic layer of forest floor	Inflow	Initial deposition (A0)
			Bark abrasion
			Stemflow
			Litterfall
		Outflow	Translocation from A0 to Soil
Soil	Surface soil of forest floor	Inflow	Translocation from A0 to Soil
		Outflow	Absorption from Soil
Bark	Bark part of trunk and branch	Inflow	Initial deposition (Bark)
			Translocation from Wood to Bark
		Outflow	Bark abrasion
			Stemflow
			Surface infiltration from Bark
Wood	Wood part of trunk and branch	Inflow	Absorption from Soil
			Surface infiltration from Bark
		Outflow	Translocation from Wood to Bark
			Translocation from Wood to Leaf
Leaf	Living leaves in tree crown	Inflow	Translocation from Wood to Leaf
		Outflow	Litterfall
(b) sugi			
A0	Organic layer of forest floor	Inflow	Initial deposition (A0)
			Bark abrasion
			Stemflow
			Litterfall
			Throughfall
		Outflow	Translocation from A0 to Soil
Soil	Surface soil of forest floor	Inflow	Translocation from A0 to Soil
		Outflow	Absorption from Soil
Bark	Bark part of trunk and branch	Inflow	Initial deposition (Bark)
			Translocation from Wood to Bark
		Outflow	Bark abrasion
			Stemflow

(continued)

Table 13.1 (continued)

Compartment	Description	Inflow/ outflow	Related parameter for translocation
			Surface infiltration (Bark)
Wood	Wood part of trunk and branch	Inflow	Absorption from Soil
			Surface infiltration from Bark
		Outflow	Translocation from Wood to Bark
			Translocation from Wood to OLI
Old leaf surface (OLS)	More than 1-year-old leaves with initial deposition of ^{137}Cs	Inflow	Initial deposition (crown)
		Outflow	Litterfall (OLS)
			Surface infiltration (leaf)
			Throughfall
Old leaf interior (OLI)	More than 1-year-old leaves excluding old leaf surface	Inflow	Surface infiltration (leaf)
			Translocation from Wood to OLI
			Aging of New leaf
		Outflow	Translocation from OLI to New leaf
			Litterfall (OLI)
New leaf	Currently growing leaves in tree crown	Inflow	Translocation from OLI to New leaf
		Outflow	Aging of New leaf

Table 13.2 Descriptions of forest sites in Ohtama and Kawauchi measured in 2011. Tree height was referred to Kajimoto et al. (2014), and the others were done to Imamura et al. (2017)

	Ohtama	Kawauchi
Main tree species	Konara	Sugi
Initial deposition; $F(0)$ (kBq m^{-2})	44	630
Stand age (y)	43	43
Tree density (n m^{-2})	0.55	0.98
DBH (cm)	17.5	18.8
Height (m)	12.3	14.3

decomposition; therefore, the amount of ^{137}Cs activity in A0 was always larger than that in Leaf. In contrast, the sugi model showed larger ^{137}Cs activity in Total leaf (New leaf + Old leaf interior (OLI) than A0 after the drastic variations ceased, because ^{137}Cs directly deposited on Old leaf surface (OLS) was assumed to be

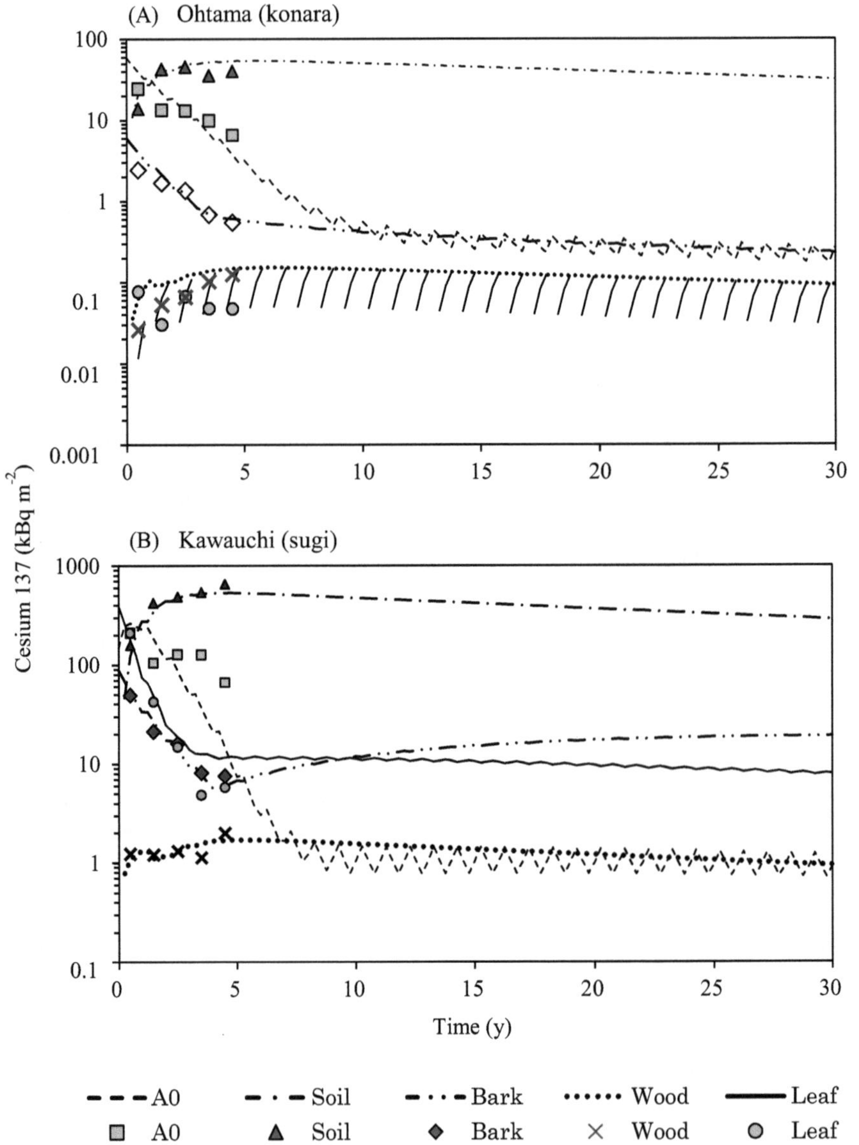

Fig. 13.2 Estimates of ^{137}Cs temporal variation for each compartment of konara forest in Ohtama and sugi forest in Kawauchi, compared to measured data. Graphs A and B show the inventory (Bq m^{-2}). In graph A, lines and symbols show estimated and measured data, respectively. The Leaf in graph B shows Total leaf (OLI + OLS + New leaf). The horizontal axis starts from winter (January to March) in 2011

disappeared until the initial 4 years, but ^{137}Cs in only one-fourth of OLI was assumed to be transferred to A0 as litter-fall every year.

13.3 Are the Sets of Best-Fit Parameters Applicable to Other Forests in Fukushima?

This model should be validated whether the estimations using the best sets of parameter values also fit well to the measured data in the other forest or not, because some of the measured data in Fig. 13.2 were used for calculating the initial parameter values. Therefore, the estimates were compared with the measured data in Yamakiya, Kawamata, Fukushima (Yamakiya).

13.3.1 Data for Comparison

Data from previous studies and originally measured data were used for comparison. A konara stand (konara A) researched by JAEA (2013) was referred to as the measured data, while, a sugi stand (sugi A) researched by JAEA (2013) and Coppin et al. (2016) were referred. And then, we collected original data in konara A and sugi A. Moreover, we collected the data of a sugi stand (sugi B) and a konara stand (konara B) in Setohachi-yama forest, Yamakiya, Kawamata, Fukushima (N: 37° 35′ 35″, E: 140° 42′ 30″; Table 13.3) to measure the data for comparison to the estimates of the konara model and the sugi one. In the konara A, the data corresponding to A0, Soil, Bark, and Wood compartments were collected ($n = 5$) in November 13, 2014 (Wang et al. 2018), and that of Leaf were collected ten times ($n = 3$ to 5) from July 25, 2011, to September 28, 2014. In the sugi A, those of New leaf and Old leaf (OLI + OLS) were collected 12 times ($n = 3$ to 5) from October 22, 2012, to May 2, 2015. In konara B and sugi B, the data sets for all of the compartments corresponding to both of the konara model (Soil, A0, Wood, Bark and Leaf) and the sugi one (Soil, A0, Wood, Bark, Old leaf and New leaf) were collected in September 24, 2015, respectively.

Table 13.3 Description of the Setohachi-yama site

	Konara	Sugi
Tree density (n m^{-2})	0.21	0.26
DBH (cm ± SD)	16.6 ± 7.3	30.0 ± 1.6
Height (m ± SD)	16.4 ± 1.9	22.7 ± 0.8
Initial deposition	774	774

13.3.2 Model Estimates Versus the Measured Data in Yamakiya

Figure 13.3 shows the ratios of ^{137}Cs in compartments estimated by the model, using the same best sets of parameter values for konara forest in Ohtama and sugi one in Kawauchi (Table 13.3), and the measured data in Yamakiya. For both the konara forests and the sugi forests, notable differences were found between the model estimates and the measured data. In both of the konara forest and the sugi one, Wood were much smaller ^{137}Cs in the estimates than in the measured data. In the konara forest, Leaf was much smaller in the estimate than in the measured data. This result means that konara and sugi incorporated and translocated higher percentages of ^{137}Cs to the initial deposition ^{137}Cs in Yamakiya than in Ohtama and Kawauchi, suggesting the existence of higher ratio of accessible ^{137}Cs in forest ecosystems in Yamakiya. Comparing the ratio of measured data, the averaged ratios of ^{137}Cs in Wood and Leaf in the konara forests were 17 times and 13 times larger in the measured data than in the estimates, respectively, while those of Bark, A0, and Soil were 2.6, 2.2, and 0.5 times larger in the measured data than in the estimates, respectively. In the sugi forests, Wood was 26 times larger in the measured data than in the estimates, while Bark, A0, Soil, and Total leaf (OLI + OLS + New leaf) were 6.1, 4.5, 2.0, and 1.5 times larger in the measured data, respectively.

Two pathways are hypothesized for the higher ^{137}Cs incorporation in trees. One is that the initial deposition in Yamakiya contained the higher ratio of accessible ^{137}Cs for plants; thus, the ^{137}Cs was incorporated by konara and sugi through surface infiltration and/or root uptake immediately after the FDNPP accident, and then, the initial incorporation of the higher ^{137}Cs have been kept in trees higher for 5 years after the FDNPP accident. Previous studies estimated that wet/dry ratio of ^{137}Cs in the initial deposition was variously distributed in Fukushima (e.g., Gonze et al. 2014; Kanasashi et al. 2016), being the higher wet/dry ratio in the sites of Yamakiya than in those of Kawauchi and Ohtama (Gonze et al. 2014). Thus, the local difference of wet/dry ratio was considered to affect the initial incorporation of ^{137}Cs for trees.

The other was the continual incorporation of higher ^{137}Cs in Yamakiya than in Ohtama and Kawauchi, mainly root uptake. The difference was considered to be attributed to the retention strength of cesium in soil (see Chap. 3) and/or mycorrhizal symbiosis (see Chap. 10). Unfortunately, the effective measured data of wood parts in 2011 or 2012 were not found for both of konara and sugi in Yamakiya; thus, this study could not mention the temporal variation of ^{137}Cs in those.

Even between the forests in Yamakiya, temporal variations of ^{137}Cs might be various between the two sites. For example, in Fig. 13.3b, in sugi A site whose samples were collected until 4 years after the FDNPP accident, ^{137}Cs ratios in Soil became larger than those in A0 dependent on time during the initial 3 years. In contrast, in sugi B site, ^{137}Cs ratios in Soil were similar to that in A0, though they were collected in 2015, 4 years after the FDNPP accident. The results suggested that ^{137}Cs flux from A0 to the Soil in sugi A was larger than that in sugi B, and/or transferring ^{137}Cs from trees to A0 was smaller in sugi A than in sugi B, respectively.

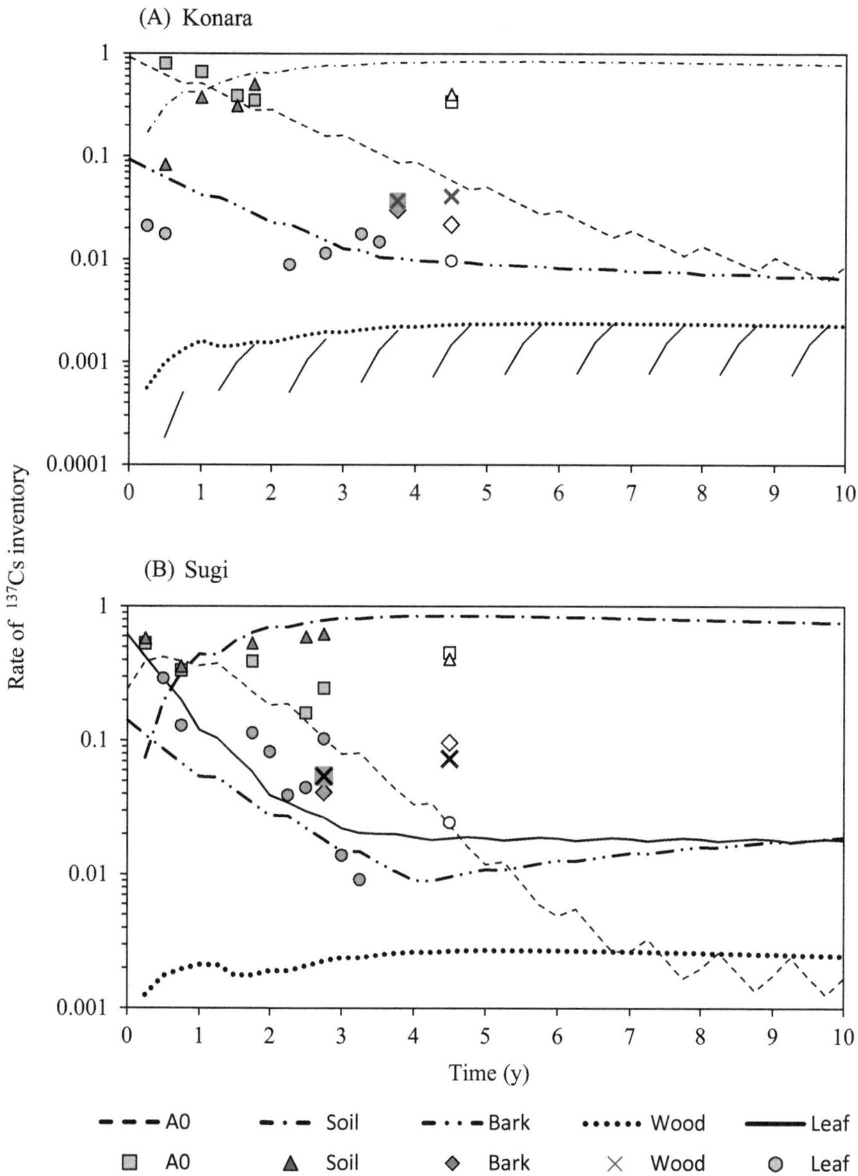

Fig. 13.3 Comparing the rates of ^{137}Cs temporal variation between the estimates and the measured data in Yamakiya. Graphs A and B show konara forest and sugi forest, respectively. As the initial deposition of ^{137}Cs was different between konara A and konara B, and between sugi A and sugi B, the measured data and the estimates are showed as rates to the initial deposition (decay-corrected at March 17, 2011). The vertical axis shows the rate of ^{137}Cs inventory and the initial deposition 1.0. The Leaf in graph B shows Total leaf (OLI + OLS + New leaf). Lines and symbols show estimated and measured data, respectively. Symbols with grey color show konara A site or sugi A site, while, those with solid color show konara B or sugi B. The horizontal axis starts from winter (January to March) in 2011

This chapter does not try to determine the optimal parameter values for the konara model and the sugi model in Yamakiya. There was no measured data of Bark and Wood until 2014 in the konara forest and until 2013 in the sugi one; therefore, the properness of parameter, *Surface infiltration of Bark* could not be validated, because surface infiltration was defined to be terminated during 2011. Similarly, stemflow was defined to be terminated until 2014; hence, there was only one measured data to validate the parameter, *stemflow*, being difficult to evaluate the optimal parameter values for both of konara forest and sugi one, properly.

The models of this study have to change the optimal set of parameter values dependent on the characteristics of local forests, one by one. Such the process is not considered to be useful for the estimation of ^{137}Cs dynamics to the various forests because ^{137}Cs concentration or amount per unit area of all the compartments in the models has to be measured in every forest. Thus, finding other effective parameters will be required to express the local variations of ^{137}Cs circulation in the forest.

13.4 Conclusion

This chapter developed a dynamic model of ^{137}Cs in forest ecosystems and provided examples of future estimation in deciduous broad-leaved (konara) forest and evergreen coniferous needle-leaved Japanese cedar (sugi) forest. This model did not take into account ^{137}Cs stored in root due to little data to apply the model, but adequate estimates of ^{137}Cs dynamics in both of the forests could be described. Quantitative data for surface infiltration of ^{137}Cs from bark and leaf are necessary because they are considered to be important factors in the early phase after the initial deposition of ^{137}Cs happened.

13.5 Method

13.5.1 How to Predict ^{137}Cs Distribution in Forest Ecosystems

13.5.1.1 Model Description

In order to the support to develop the dynamic model, previous studies of Schell et al. (1996), Thiry et al. (2018) and Kanasashi et al. (unpublished) were applied. This model was composed of compartments representing organic layer and soil in the forest floor, and tree parts of konara and sugi. In this model, ^{137}Cs circulated among the compartments, with ^{137}Cs input from the initial deposition derived from the FDNPP reactors, and without any output from the forest due to small ^{137}Cs fluxes from forests to other areas (Ueda et al. 2013; Shinomiya et al. 2014; Yoshimura et al.

2015). The loss of ^{137}Cs from the forest was assumed by only decay correction from March 17, 2011. The initial deposition of ^{137}Cs (Bq m^{-2}) was determined using the fourth airborne gamma-ray monitoring survey data (Japanese Ministry of Education, Culture, Sports, Science and Technology 2011). Monitoring seasons of reference data for the model were different; thus, this model added parameters of seasonal change in ^{137}Cs (Kanasashi et al. unpublished). A season means a quarter of a year: spring (April to June), summer (July to September), autumn (October to December), or winter (January to March).

And then, four steps were conducted for the model development as follows.

1. Defining the compartments in forest ecosystems and the directions of ^{137}Cs flow.
2. Applying equations to represent the ^{137}Cs flow quantitatively.
3. Defining the initial parameters of the ^{137}Cs flows between the compartments.
4. Tuning the initial parameters to be optimal to the measured data.

13.5.2 The First Step: Defining the Compartments in Forest Ecosystems and the Directions of ^{137}Cs Transfer

Figure 13.1 is a conceptual description of a compartment model of ^{137}Cs cycling in the konara forest and the sugi forest.

We simply defined organic layer (A0) and soil (Soil) in abiotic compartments, because most of the ^{137}Cs in soil was deposited in surface soil for a long time, according to the Chernobyl accident (Yoshida et al. 2004); thus, this model did not consider ^{137}Cs migrant among different soil layers.

Of the compartments, Wood refers to the inner portions of trunks and branches, and Bark refers to the surfaces of these tree parts. In Table 13.1b, the OLS compartment refers to the parts of needle leaves on which ^{137}Cs was directly deposited immediately after the FDNPP accident. All the compartments and the parameters in Fig. 13.1 were described in Table 13.1 and 13.4, respectively.

13.5.3 The Second Step: Applying Equations to Represent the ^{137}Cs Flow Quantitatively

For the second step, the mathematical model was expressed as the sum of ^{137}Cs flow input from other compartments and output to another compartment. Linear equations were selected to calculate the transfer of ^{137}Cs between compartments, taken from Schell et al. (1996):

Table 13.4 Parameters for ^{137}Cs transferring of models with initial values used in the Monte Carlo test. The parameters with two initial values due to more than two different reference data were shown: the lowest one was above, and the highest one was below

Parameter	Symbol	Initial value		Source
		Konara	Sugi	
Initial deposition (A0)	k_α	0.77	0.15	Kato et al. (2017)
Initial deposition (Bark)	k_β	0.23	0.15	Kato et al. (2017)
Initial deposition (OLS)	k_γ	–	0.70	Kato et al. (2017)
A0 to Soil	v_{12}	0.10	0.12	JAEA (2013)
		0.18	0.28	Kato et al. (2017)
Root uptake	v_{23}	0.0022	0.0027	Imamura et al. (2017)
Translocation from Wood to Bark	v_{34}	0.43	0.19	Ohashi et al. (2014) and Coppin et al. (2016)
Translocation from Wood to Leaf	v_{35}	0.37	–	Imamura et al. (2017)
Translocation from Wood to OLI	v_{36}	–	0.20	Imamura et al. (2017)
			0.42	
Bark abrasion	v_{41}	0.0077	0.0030	Endo et al. (2015)
		0.049	0.0090	
Stemflow	v_{41}'	0.31	0.22	Imamura et al. (2017)
Surface infiltration of Bark	v_{43}	0.01	0.01	Arbitrary
Litterfall (konara)	v_{51}	1.0	–	Arbitrary
Translocation from OLI to New leaf	v_{67}	–	0.20	Arbitrary
Litterfall from OLI	v_{61}	–	0.125	Miyaura et al. (1995)
Litterfall from OLS	v_{81}	–	0.125	Arbitrary
Throughfall	v_{81}'	–	0.34	Imamura et al. (2017)
Translocation from New leaf to OLI	v_{76}	–	1.0	Arbitrary
Surface infiltration of OLS	v_{86}	–	0.089	Nishikiori et al. (2015)
			0.21	

$$\frac{dQ_i(t)}{dt} = \sum_{j;i\neq j} v_{ji} \bullet Q_j - \sum_{n;i\neq n} v_{in} \bullet Q_i + F_i(t) - \lambda \bullet Q_i \qquad (13.1)$$

where $Q_i\ (t)$ represents the total radionuclide activity stored in compartment i and time t of the model, v_{ji} represents the rate constant for the ^{137}Cs flow per time from the j-th compartment to the i-th compartment, v_{in} represents the rate constant for the flow from the i-th compartment to the n-th compartment, $F_i\ (t)$ represents the external source term for the i-th compartment (Bq m^{-2} × time^{-1}), and λ is the radionuclide decay rate.

Based on Eq. (13.1), two types of parameters were incorporated, which controlled the strength of the flows and seasonal change of the flows because some of the referred data were monitored at a different season. Therefore, Eq. (13.1) was modified as follows:

$$\frac{dQ_i(t)}{dt} = \sum_{j;i\neq j} s_{ji} \bullet v_{ji} \bullet Q_j - \sum_{n;i\neq n} s_{in} \bullet v_{in} \bullet Q_i + k \bullet F_i(0) - \lambda \bullet Q_i \tag{13.2}$$

where t represents the unit time of a season (a quarter of year), and s_{ji} and s_{in} represent the seasonal change (time-dependent rates) for the flow from the j-th compartment to the i-th compartment and from the i-th compartment to the n-th compartment, respectively. The parameter k represents the ratio of the initial deposition and controls the allocation of that on A0, bark, and canopy, but the deposition on leaves was caused at the compartment of OLS of sugi (Nishikiori et al. 2015; Kanasashi et al. 2015) and not caused at the leaf of konara, due to no leaf at the dormant season when the FDNPP accident took place. Then, F_i (0) means the amount of initial deposition; thus, time is only zero ($t = 0$). The model must satisfy that the variables are positive under the following ten equations (Schell et al. 1996):

$$\frac{dQ_1(t)}{dt} = k_\alpha \bullet F(0) + (s_{41} \bullet v_{41} + s'_{41} \bullet v_{41}') \bullet Q_4 + s_{51} \bullet v_{51} \bullet Q_5 - s_{12} \bullet v_{12} \bullet Q_1 - \lambda \bullet Q_1 \tag{13.3}$$

$$\frac{dQ_1(t)}{dt} = k_\alpha \bullet F(0) + (s_{41} \bullet v_{41} + s'_{41} \bullet v_{41}') \bullet Q_4 + (s_{61} \bullet v_{61}) \bullet Q_6 + (s_{81} \bullet v_{81} + s'_{81} \bullet v_{81}') \bullet Q_8 - s_{12} \bullet v_{12} \bullet Q_1 - \lambda \bullet Q_1 \tag{13.3$'$}$$

$$\frac{dQ_2(t)}{dt} = s_{12} \bullet v_{12} \bullet Q_1 - s_{23} \bullet v_{23} \bullet Q_2 - \lambda \bullet Q_2 \tag{13.4}$$

$$\frac{dQ_3(t)}{dt} = s_{23} \bullet v_{23} \bullet Q_2 + s_{43} \bullet v_{43} \bullet Q_4 - (s_{34} \bullet v_{34} + s_{35} \bullet v_{35}) \bullet Q_3 - \lambda \bullet Q_3 \tag{13.5}$$

$$\frac{dQ_3(t)}{dt} = s_{23} \bullet v_{23} \bullet Q_2 + s_{43} \bullet v_{43} \bullet Q_4 - (s_{34} \bullet v_{34} + s_{36} \bullet v_{36}) \bullet Q_3 - \lambda \bullet Q_3 \tag{13.5$'$}$$

$$\frac{dQ_4(t)}{dt} = k_\beta \bullet F(0) + s_{34} \bullet v_{34} \bullet Q_3 - (s_{41} \bullet v_{41} + s'_{41} \bullet v_{41}' + s_{43} \bullet v_{43}) \bullet Q4 - \lambda \bullet Q_4 \tag{13.6}$$

$$\frac{dQ_5(t)}{dt} = s_{35} \bullet v_{35} \bullet Q_3 - s_{51} \bullet v_{51} \bullet Q_5 - \lambda \bullet Q_5 \tag{13.7}$$

$$\frac{dQ_6(t)}{dt} = s_{36} \bullet v_{36} \bullet Q_3 + s_{76} \bullet v_{76} \bullet Q_7 + s_{86} \bullet v_{86} \bullet Q_8 - (s_{61} \bullet v_{61} + s_{67} \bullet v_{67}) \bullet Q_6 - \lambda \bullet Q_6 \tag{13.8}$$

$$\frac{dQ_7(t)}{dt} = s_{67} \bullet v_{67} \bullet Q_6 - s_{76} \bullet v_{76} \bullet Q_7 - \lambda \bullet Q_7 \quad (13.9)$$

$$\frac{dQ_8(t)}{dt} = k_\gamma \bullet F(0) - (s_{86} \bullet v_{86} + s_{81} \bullet v_{81} + s'_{81} \bullet v_{81}') \bullet Q_8 - \lambda \bullet Q_8 \quad (13.10)$$

where,

$Q_1 =$ ^{137}Cs radioactivity in A0 (Bq m^{-2});
$Q_2 =$ ^{137}Cs radioactivity in Soil (Bq m^{-2});
$Q_3 =$ ^{137}Cs radioactivity in Wood (Bq m^{-2});
$Q_4 =$ ^{137}Cs radioactivity in Bark (Bq m^{-2});
$Q_5 =$ ^{137}Cs radioactivity in Leaf of konara (Bq m^{-2});
$Q_6 =$ ^{137}Cs radioactivity in OLI of sugi (Bq m^{-2});
$Q_7 =$ ^{137}Cs radioactivity in New leaf of sugi (Bq m^{-2});
$Q_8 =$ ^{137}Cs radioactivity in OLS of sugi (Bq m^{-2});

Then, s_{41} and v_{41} represent parameters for *Bark abrasion,* s_{41}' and v_{41}' represent those for *Stemflow*, s_{81} and v_{81} represent parameters for *Litterfall from OLS*, s_{81}' and v_{81}' represent those for *Throughfall*, respectively (Table 13.5). The parameters k_α, k_β, and k_γ are those for the allocation of the *Initial deposition on A0, Bark*, and *OLS*, respectively.

Equations (13.4) and (13.6) were used in both the konara model and the sugi one, Eqs. (13.3), (13.5), and (13.7) were used only for the konara model, and Eqs. (13.3′), (13.5′), (13.8), (13.9), and (13.10) were used only for the sugi model.

13.5.4 The Third Step: Defining the Initial Parameters of the ^{137}Cs Flows Between the Compartments

To determine the parameter values v_{ij} of the mathematical model Eqs. (13.3) to (13.10), the initial parameter values were adopted or calculated from previous studies. And then, the initial parameters were tuned to optimally fit measured data set (explained at the next step). The mathematical model equations had another parameter s_{ij}, controlling the time-dependent rates, but the parameter was fixed and not covered from the tuning because s_{ij} was only zero or one, basically (mentioned later).

The initial parameter values v_{ij} were taken from previous studies for the FDNPP accident (Table 13.4). Methods to determine the initial parameter values are discussed in the following sections.

Table 13.5 Parameters for seasonal change of ^{137}Cs transferring of models. These parameter values changed among unit time (season), one (transferring ^{137}Cs) or zero (non-transferring ^{137}Cs), but *Litterfall from OLI* and *Litterfall from OLS* are different parameter values, and then, some of the parameters were defined that their values were changed zero constantly from 1 to 4 years after the FDNPP accident (see Sect. 13.5.4.11). Some of the parameter values and their time variations controlling both of the konara model and sugi one were equal. Sp, spring; Su, summer; Au, autumn; and Wi, winter

Parameter	Symbol	Seasonal variation	
		Value = 1	Value = 0
A0 to Soil	s_{12}	Sp, Su, Au	Wi
Root uptake	s_{23}	Sp, Su, Au	Wi
Translocation from Wood to Bark	s_{34}	Sp, Su, Au	Wi
Translocation from Wood to Leaf	s_{35}	Sp, Su, Au	Wi
Translocation from Wood to OLI	s_{36}	Sp, Su, Au	Wi
Bark abrasion	s_{41}	All seasons	None
Stemflow	s_{41}'	Sp, Su, Au from 2011 to 2014	Wi from 2011 to 2014 and after 2015
Surface infiltration of Bark	s_{43}	Sp, Su, Au in 2011	After Wi in 2011
Litterfall (konara)	s_{51}	Au	Sp, Su, Wi
Translocation from OLI to New leaf	s_{67}	Sp, Su, Au	Wi
Litterfall from OLI	s_{61}	Au, Wi	Sp, Su
Litterfall from OLS	s_{81}	Au, Wi from 2011 to 2014	Sp, Su from 2011 to 2014 and after 2015
Throughfall	s_{81}'	Sp, Su, Au from 2011 to 2014	Wi from 2011 to 2014 and after 2015
Translocation from New leaf to OLI	s_{76}	Sp	Su, Au, Wi
Surface infiltration of OLS	s_{86}	Sp, Su, Au in 2011	After Wi in 2011

13.5.4.1 Initial Deposition Ratios

Cesium 137 in the initial fallout was defined to be deposited on OLS, Bark, and A0 in the sugi model, due to the dormant season and no New leaf in 2011 in the crown when the FDNPP accident happened. For the konara model, Bark and A0 were contaminated by the ^{137}Cs in the initial deposition, due to no leaf as the same reason. Their initial parameter values were estimated using the data of Kato et al. (2017). The initial parameters of A0 (k_α), Bark (k_β), OLS (k_γ) for the sugi model were 0.15, 0.15, and 0.70, respectively. And then, those of A0 (k_α) and Bark (k_β) for the konara model were 0.77 and 0.23, respectively. The sum of the parameter values ($k_\alpha + k_\beta + k_\gamma$ for the sugi model; $k_\alpha + k_\beta$ for the konara model) is one.

13.5.4.2 Bark Abrasion

The initial parameter of *Bark abrasion* (v_{41}) was defined by Endo et al. (2015), using the flux data of ^{137}Cs with bark-litterfall. From the parameter values of *the initial deposition* (bark), the parameter values v_{41} were calculated. The literature had the range of initial deposition data; thus, the initial parameter values v_{41} were also ranged from 0.0077 to 0.049 of konara (calculated from two forest stands) and from 0.0030 to 0.0090 of sugi.

13.5.4.3 Throughfall and Stemflow

In this model, *Throughfall* was defined to be derived from only OLS. The initial parameter values of *Throughfall* (v_{81}') and *Stemflow* (v_{41}') were calculated from the data provided by Imamura et al. (2017). The difference between needle leaf ^{137}Cs in 2011 and the initial deposition ^{137}Cs on the needle leaf surface was calculated, and then, the ratio between the difference and the initial deposition ^{137}Cs was defined as the initial parameter value ($v_{81}' = 0.34$).

Stemflow for konara and sugi was also calculated using the differences of ^{137}Cs data between bark in 2011 and the initial deposition ^{137}Cs on bark. Then, v_{41}' of konara and sugi were 0.31 and 0.22, respectively.

13.5.4.4 A0 to Soil

For calculating the initial parameter (v_{12}), ^{137}Cs flux from A0 to soil was calculated using the following equation:

$$\text{Flux}_{\rightarrow \text{Soil}}(t) = Q_1(t) + \text{Flux}_{\rightarrow A0}(t) - Q_1(t-1) \tag{13.11}$$

where $\text{Flux}_{\rightarrow \text{Soil}}(t)$ was ^{137}Cs flux from A0 to Soil at season *t*, $\text{Flux}_{\rightarrow A0}(t)$ was the ^{137}Cs flux to A0 through sum of stemflow, throughfall, and litterfall. And then, the ratios between ^{137}Cs of $\text{Flux}_{\rightarrow \text{Soil}}(t)$ and sum of ^{137}Cs of $\text{Flux}_{\rightarrow \text{Soil}}(t)$ and total of $Q_1(t)$ was calculated as the value of $v_{12.}$ The A0 data was obtained from Japan Atomic Energy Agency (JAEA) (2013) and the flux data was from Kato et al. (2017). The parameter values of v_{12} for the konara model and the sugi model were ranged from 0.10 to 0.18 and 0.12 to 0.28 (calculated from two forest stands), respectively.

13.5.4.5 Surface Infiltration from OLS and Bark

Nishikiori et al. (2015) reported that ^{137}Cs inside the leaves of evergreen trees accounted for 17–38% of all ^{137}Cs in leaves within 6 months after the accident. Surface infiltration of ^{137}Cs was dominant during the early phase of initial deposition, compared with the root uptake of ^{137}Cs (Rantavaara et al. 2012). Therefore, we assumed that most of the ^{137}Cs inside the leaves was absorbed from the surface of the leaves immediately after the FDNPP accident, and the *surface infiltration from OLS* was assumed to occur during the initial 3 seasons and the parameter v_{86} for the only sugi model were calculated as 0.089–0.21.

Surface infiltration of ^{137}Cs from the bark was considered as one of the ^{137}Cs pathways at the FDNPP accident (Wang et al. 2016; Wang et al. 2018), but quantitative data were not obtained. Generally, the bark prevents water loss from inner parts of the wood, but water uptake through bark has been suggested to occur under a specific circumstance (Mason et al. 2015). Therefore, low parameter values for *surface infiltration of Bark* ($v_{43} = 0.01$) were arbitrarily chosen in the konara and sugi models.

13.5.4.6 Root Uptake

The initial parameter of *Root uptake* (v_{23}) was calculated by the ratio of ^{137}Cs (Bq m^{-2}) in wood and in soil using the data of Imamura et al. (2017), including data gathered over several years. And then, v_{23} of the konara model and the sugi model was set to 0.0022 and 0.0027, respectively.

13.5.4.7 Litterfall

Konara, which is deciduous broad-leaved tree, sheds all leaves in autumn, thus the initial parameter of *litterfall* (v_{51}) of the konara model was 1.0 in every autumn. In contrast, sugi, which is an evergreen needle-leaved tree, sheds some fraction of needle leaves, except for current-year ones. As sugi needle leaves typically undergo senescence and fall within 5 years (Miyaura et al. 1995), we hypothesized that a quarter of the ^{137}Cs in the OLD leaf compartment (OLS + OLI) was transported to A0 each year due to senescence and litterfall. And then, the senescent needle leaves of sugi mainly fall during two seasons: autumn and winter (Katagiri et al. 1990; Yoshida 2006; Kanasashi and Hattori 2011). Therefore, v_{61} and v_{81} for the sugi model was set to 0.125.

13.5.4.8 Translocation from Wood to Leaf for Konara Model and Wood to OLI for Sugi Model

Using the data of ^{137}Cs (Bq m^{-2}) in wood and leaves of Imamura et al. (2017), the initial parameters of *translocation from Wood to Leaf* for the konara model (v_{35}) and *Wood to OLI* for the sugi model (v_{36}) were calculated. For konara, all of the ^{137}Cs in Leaf was assumed to be translocated from Wood; thus, the rate of ^{137}Cs in leaves and wood referred from the literature was calculated as $v_{35} = 0.37$. On the other hand, old needle leaves of sugi contained ^{137}Cs which had already translocated and stored after the FDNPP accident, and was translocated from wood in current year because sugi is evergreen tree; moreover, an affection of the surface infiltration should be taken into consideration. Therefore, the initial parameter v_{36} was calculated as the following assumption: from Imamura et al. (2017), the data of needle leaves collected in 2015 were defined no affection of the adhered ^{137}Cs. And then, one-fifth of the ^{137}Cs inventory (Bq m^{-2}) in the needle leaves was defined as ^{137}Cs in current-year needle leaves, and to be equal to annual ^{137}Cs flux from Wood to OLI. As this calculation, v_{36} was ranged from 0.20 to 0.42 (calculated from two forest stands).

13.5.4.9 Translocation from OLI to New Leaf and New Leaf to OLI for Sugi Model

As mentioned above, one-fifth of the ^{137}Cs (Bq m^{-2}) in the needle leaves was assumed to be equal to ^{137}Cs translocated from Wood to OLI. Therefore, the initial parameter from OLI to New leaf (v_{67}) was calculated as the rate of the defined ^{137}Cs between current-year needle leaves and old needle leaves, thus v_{67} was 0.20.

As current-year leaves become one-year-old leaves in every next spring, the initial parameter *translocation from New leaf to OLI* (v_{76}) was 1.0.

13.5.4.10 Translocation from Wood to Bark

Bark also deposited ^{137}Cs with high activity on the surface and was obscured ^{137}Cs translocation from Wood to Bark. Therefore, the initial parameter v_{34} was calculated as the ratio of ^{137}Cs (Bq m^{-2}) between inner bark and wood. For this calculation, Ohashi et al. (2014) and Coppin et al. (2016) were referred for the initial parameters of the konara ($v_{34} = 0.43$) model and the sugi model ($v_{34} = 0.19$), respectively.

13.5.4.11 Seasonal Changes (No-Tuning)

As Eqs. (13.3) to (13.10), this model contains the parameter to control the seasonal changes of ^{137}Cs flow between the compartments, because, for example, the model was defined that physiological uptakes and translocations of radiocesium in konara and sugi trees did not occur in winter due to the dormant season, and radiocesium transport from the A0 layer to the soil also did not in winter due to snow cover on the forest floor. Furthermore, some of the previous studies referred to calculate the initial parameters measured ^{137}Cs in different seasons. Therefore, the parameters for controlling the seasonal change were organized in this model.

The parameter was always zero or one, except for *Litterfall from OLI* (s_{61}) and *Litterfall from OLS* (s_{81}). And then, the parameter values of s_{ij} and s_{in} were fixed and not intended to tune the initial parameters at the next step. This model defined the values as follows:

- All of ^{137}Cs in Wood for the konara model is always fallen in autumn, so the parameter of *Litterfall* s_{51} is one in the season ($t = 3, 7, 11...$) and zero in the others.
- *Litterfall from OLI* (s_{61}) was 0.5, when the season was every autumn ($t = 3, 7, 11...$) and every winter ($t = 4, 8, 12...$), respectively, because sugi leaves fall during the seasons (Katagiri et al. 1990; Yoshida 2006; Kanasashi and Hattori 2011).
- *Litterfall from OLS* (s_{81}) had to be terminated until 4 years ($t = 16$) after the FDNPP accident because sugi leaves typically undergo senescence within 5 years (Miyaura et al. 1995). In March 2011, current-year needle leaves did not start growing, so ^{137}Cs in the initial deposition was adhered on from 1-year needle leaves to 4-year ones, and then, ^{137}Cs in OLS was assumed to be constantly transferred to A0 through litterfall in autumn and winter like OLI. Therefore, s_{81} was defined that, in autumn ($t = 3, 7, 11$ and 15) and winter ($t = 4, 8, 12$ and 16) until 4 years ($t = 16$), s_{81} varied 1/8, 1/7, 1/6, 1/5, 1/4, 1/3, 1/2, and 1 dependent on increasing t.
- *Surface infiltration of OLS* (s_{86}) in the sugi model and *Surface infiltration of Bark* (s_{43}) in the konara model and the sugi model were one from spring to autumn, but this model was defined that these surface infiltration of ^{137}Cs had continued only for three seasons in 2011 ($t = 1$ to 3), so the others were zero.
- *Throughfall* (s_{81}') of sugi, and *Stemflow* (s_{41}') of konara and sugi were zero in winter ($t = 4, 8, 12...$) and one in the other seasons, and then, both of the parameters were zero in all seasons 4 years after the FDNPP accident because most of the ^{137}Cs transferring by throughfall was largely decreased until 4 years ($t = 16$) after the FDNPP accident (Kato et al. 2017).
- *Bark abrasion* (s_{41}) was always one due to a physical phenomenon.
- *Translocation from New leaf* to *OLI* (s_{76}) was one in only spring ($t = 1, 5, 9...$), because New leaf becomes one-year older in next spring every year.
- Other parameters (s_{12}, s_{23}, s_{34}, s_{35}, s_{36}, and s_{67}) were zero only in winter ($t = 4, 8, 12...$) and one in the other seasons (see Table 13.5).

13.5.5 The Fourth Step: Tuning the Initial Parameters to Be Optimal to Measured Data

13.5.5.1 Best Set of Parameter Values

To find the best set of parameter values, the Monte Carlo method was used (Wang et al. 2012), repeating the output results of the model by changing 100,000 sets of parameter values applied using random numbers. Only parameters v_{ij} k_α, k_β, and k_γ, to control the flow strengths at the Eq. (13.3) to (13.10) were tuned (Table 13.6). The time-dependent rates s_{ji} were not tuned because we defined that physical and physiological seasonal changes followed as the previous section 13.5.4.11.

Table 13.6 Optimized parameter values in Ohtama of the konara model and Kawauchi of the sugi model

Parameter	Symbol	Calibrated initial value	
		Konara	Sugi
Initial deposition (A0)	k_α	0.91	0.24
Initial deposition (Bark)	k_β	0.09	0.14
Initial deposition (OLS)	k_γ	–	0.62
A0 to Soil	v_{12}	0.19	0.31
Root uptake	v_{23}	0.0014	0.0016
Translocation from Wood to Bark	v_{34}	0.23	0.028
Translocation from Wood to Leaf	v_{35}	0.33	–
Translocation from Wood to OLI	v_{36}	–	0.21
Bark abrasion	v_{41}	0.046	0.0042
Stemflow	v_{41}'	0.13	0.20
Surface infiltration of Bark	v_{43}	0.007	0.0090
Litterfall (konara)	v_{51}	1.0	–
Translocation from OLI to New leaf	v_{67}	–	0.20
Litterfall from OLI	v_{61}	–	0.125
Litterfall from OLS	v_{81}	–	0.125
Throughfall	v_{81}'	–	0.32
Translocation from New leaf to OLI	v_{67}	–	1.0
Surface infiltration of Leaf	v_{86}	–	0.016

For the Monte Carlo method, the initial parameter values in Table 13.4 was randomly changed by ±75%. Some of the initial parameters with two different values calculated from two sites were provided ranges using their largest values +75% and their smallest values −75% for creating random numbers for the method. To find the best set of parameter values, in which a set of ^{137}Cs values in the compartments best fits a set of measured data, the differences between total inventory of ^{137}Cs in compartments in the *t*-th season and the total ^{137}Cs inventory of the measured data in the same season should be taken into consideration. Total ^{137}Cs inventories in the *t*-th season were expected to be equal between that of model estimation and that of the measured data, or that of model estimation is a little larger than that of the measured data, because, for example, this model defined no ^{137}Cs transfer from surface soil to deeper one. However, the differences of total ^{137}Cs inventories between the measured data and the model estimation in some seasons are non-negligible. The differences are attributed to the fact that large uncertainties of ^{137}Cs deposition in airborne monitoring (Torii et al. 2012) and heterogeneous ^{137}Cs distribution within the same compartment in a forest. Therefore, we defined the best set of parameters as that where, in season *t*, the ratio of ^{137}Cs in the *i*-th compartment to the total ^{137}Cs of all compartments is closest to the ratio of measured ^{137}Cs for the *i*-th compartment to the total measured ^{137}Cs for all compartments. This definition is expressed by the following function:

$$D^2 = \sum_{i=1}^{n} \frac{(M-E)^2}{E}, M = \left(\frac{m_i(t)}{m_{\text{all}}(t)}\right), E = \left(\frac{e_i(t)}{e_{\text{all}}(t)}\right) \quad (13.12)$$

where $m_i(t)$ and $m_{\text{all}}(t)$ are measured data and total ^{137}Cs of all the measured data in season *t*, respectively, corresponding to $e_i\ (t)$, and $e_{\text{all}}\ (t)$, which are the ^{137}Cs in the *i*-th compartment of the models in season *t* and total ^{137}Cs of all compartments in season *t*, respectively. D^2 indicates a χ-square value between the ratio of ^{137}Cs in a single/total compartment in the model and that in the measured data. The observed data was referred from Imamura et al. (2017): the konara forest in Ohtama and the sugi forest in Kawauchi (see site descriptions in Table 13.2), but the data do not separate needle leaves to three fractions (OLI, OLS and New leaf) as this model. Thus, the total of the estimated data in the three fractions (Q_6, Q_7, and Q_8) was applied for this test.

References

Coppin F, Hurtevent P, Loffredo N, Simonucci C, Julien A, Gonze MA, Nanba K, Onda Y, Thiry Y (2016) Radiocaesium partitioning in Japanese cedar forests following the "early" phase of Fukushima fallout redistribution. Sci Rep 6:37618

Endo I, Ohte N, Iseda K, Tanoi K, Hirose A, Kobayashi NI, Murakami M, Tokuchi N, Ohashi M (2015) Estimation of radioactive 137-cesium transportation by litterfall, stemflow and throughfall in the forests of Fukushima. J Environ Radioact 149:176–185

Gonze M, Renaud P, Korsakissok I, Kato H, Hinton T, Mourlon C, Simon-Cornu M (2014) Assessment of dry and wet atmospheric deposits of radioactive aerosols: application to Fukushima radiocaesium fallout. Environ Sci Technol 48:11268–11276

Hashimoto S, Matsuura T, Nanko K, Linkov I, Shaw G, Kaneko S (2013) Predicted spatio-temporal dynamics of radiocesium deposited onto forests following the Fukushima nuclear accident. Sci Rep 3:2564. https://doi.org/10.1038/srep02564

Imamura N, Komatsu M, Ohashi S, Hashimoto S, Kajimoto T, Kaneko S, Takano T (2017) Temporal changes in the radiocesium distribution in forests over the five years after the Fukushima Daiichi nuclear power plant accident. Sci Rep 7:8179

International Atomic Energy Agency (IAEA) (2002) Modeling the migration and accumulation of radionuclides in forest ecosystems. Report of the Forest Working Group of the Biosphere Modelling and Assessment (BIOMASS) Programme, Theme 3

Japan Atomic Energy Agency (JAEA) (2013). Report for establishment of methods to elucidate long-term affections of radionuclides due to Fukusima Dai-ichi Nuclear Power Plant accident. https://fukushima.jaea.go.jp/initiatives/cat03/entry06.html. Accessed in May 2015 (in Japanese)

Japanese Ministry of Education, Culture, Sports, Science and Technology (MEXT) (2011) Results of the Fourth Airborne Monitoring Survey by MEXT 2011. http://radioactivity.nsr.go.jp/ja/contents/5000/4901/24/1910_1216.pdf. Accessed in June 2016

Kajimoto T, Takano T, Saito S, Kuroda K, Fujiwara T, Komatsu M, Kawasaki T, Ohashi S, Kiyono Y (2014) Methods for assessing the spatial distribution and dynamics of radiocesium in tree components in forest ecosystems. Bull FFPRI 13:113–136

Kanasashi T, Hattori S (2011) Seasonal variation in leaf-litter input and leaf dispersal distances to streams: the effect of converting broadleaf riparian zones to conifer plantations in central Japan. Hydrobiologia 661:145–161

Kanasashi T, Sugiura Y, Takenaka C, Hijii N, Umemura M (2015) Radiocesium distribution in sugi (*Cryptomeria japonica*) in eastern Japan: translocation from needles to pollen. J Environ Radioact 139:398–406

Kanasashi T, Takenaka C, Sugiura Y (2016) Inferring the chemical form of ^{137}Cs deposited by the Fukushima Dai-ichi nuclear power plant accident by measuring ^{137}Cs incorporated into needle leaves and male cones of Japanese cedar trees. Sci Total Environ 553:643–649

Katagiri S, Kaneko N, Obatake Y. (1990) Nutrient cycling in a Sugi (*Cryptomeria japonica* D. Don) stand with insufficient management—nutrient accumulation in aboveground and soil and nutrient return by litterfall and rainfall. Bull Fac Agr, Shimane Univ, 24, 21–27 (in Japanese)

Kato H, Onda Y, Hisadome K, Loffredo N, Kawamori A (2017) Temporal changes in radiocesium deposition in various forest stands following the Fukushima Dai-ichi Nuclear Power Plant accident. J Environ Radioact 166:449–457

Mason EJ, Sperling O, Silva LC, McElrone AJ, Brodersen CR, North MP, Zwieniecki MA (2015) Bark water uptake promotes localized hydraulic recovery in coastal redwood crown. Plant, Cell Environ 39:320–328

Miyaura M, Hagihara A, Hozumi K. (1995) Estimation of gross production in a *Cryptomeria japonica* plantation on the basis of Monsi-Saeki's model of canopy photosynthesis. Bull Nagoya Univ For, 14, 49–88 (in Japanese)

Nishikiori T, Watanabe M, Koshikawa MK, Takamatsu T, Ishii Y, Ito S, Takenaka A, Watanabe K, Hayashi S (2015) Uptake and translocation of radiocesium in cedar leaves following the Fukushima nuclear accident. Sci Total Environ 502:611–616

Nishina K, Hayashi S (2015) Modeling radionuclide Cs and C dynamics in an artificial forest ecosystem in Japan –FoRothCs ver1.0. Front Environ Sci 3(61)

Ohashi S, Okada N, Tanaka A, Nakai W, Takano S (2014) Radial and vertical distributions of radiocesium in tree stems of *Pinus densiflora* and *Quercus serrata* 1.5 y after the Fukushima nuclear disaster. J Environ Radioact 134:54–60

Rantavaara A, Vetikko V, Raitio H, Aro L (2012) Seasonal variation of the ^{137}Cs level and its relationship with potassium and carbon levels in conifer needles. Sci Total Environ 441:194–208

Schell WR, Linkov I, Myttenaere C, Morel B (1996) A dynamic model for evaluating radionuclide distribution in forests from nuclear accidents. Health Phys 70:318–335

Shinomiya Y, Tamai K, Kobayashi M, Ohnuki Y, Shimizu T, Iida S, Nobuhiro T, Sawano S, Tuboyama Y, Hiruta T (2014) Radioactive cesium discharge in stream water from a small watershed in forested headwaters during a typhoon flood event. Soil Sci Plant Nutr 60:765–771

Thiry Y, Albrecht A, Tanaka T (2018) Development and assessment of a simple ecological model (TRIPS) for forests contaminated by radiocesium fallout. J Environ Radioact:190–191

Torii T, SanadaY ST, Kondo A, Shikaze Y, Takahashi M, Ishida M, Nishizawa Y, Urabe Y (2012) Investigation of radionuclide distribution using aircraft for surrounding environmental survey from Fukushima Dai-ichi nuclear power plant. JAEA-Technol:2012–2036

Ueda S, Hasegawa H, Kakiuchi H, Akata N, Ohtsuka Y, Hisamatsu S (2013) Fluvial discharge of radiocaesium from watersheds contaminated by the FukushimaDai-ichi nuclear power plant accident, Japan. J Environ Radioact 118:96–104

Wang YC, Lin YP, Huang CW, Chiang LC, Chu HJ, Ou WS (2012) A system dynamic model and sensitivity analysis for simulating domestic pollution removal in a free-water surface constructed wetland. Water Air Soil Pollut 223:2719–2742

Wang W, Hanai Y, Takenaka C, Tomioka R, Iizuka K, Ozawa H (2016) Cesium absorption through bark of Japanese cedar (*Cryptomeria japonica*). J For Res 21:251–258

Wang W, Takenaka C, Tomioka R, Kanasashi T (2018) Absorption and translocation of cesium through Konara oak (*Quercus serrata*) bark. J For Res 23:35–40

Yoshida T (2006) Spatiotemporal structure of resource dynamics and microarthropod communities in detrital food webs in a Cryptomeria japonica forest. Nagoya Univ For Sci 25:107–153. in Japanese

Yoshida S, Muramatsu Y, Dvornik AM, Zhuchenko TA, Linkov I (2004) Equilibrium of radiocesium with stable cesium within the biological cycle of contaminated forest ecosystems. J Environ Radioact 75:301–313

Yoshimura K, Onda Y, Kato H (2015) Evaluation of radiocaesium wash-off by soil erosion from various land uses using USLE plots. J Environ Radioact 139:362–369

Chapter 14
Future Perspective

Nobuhiro Kaneko, Tatsuhiro Ohkubo, Naoki Hijii, and Chisato Takenaka

Abstract Radiocesium dynamics in forests contaminated after Fukushima Daiichi Nuclear Power Plant (FDNPP) accident was studied during the early phase (ca. until 5 years). Despite the difference in climate and forest types, the major dynamics of radiocesium followed the case of Chernobyl Nuclear Power Plant (ChNPP). Because there were rather few observations of deformity of organisms, the radiocesium dynamics in forest ecosystem was organized by ecological interaction based on physicochemical cycling of elements in forest ecosystem. Biological differences of radiocesium translocation within plant bodies in canopy tree species were clearly shown by the detailed studies on physiology and ecology of trees, especially Japanese red cedar (*Cryptomeria japonica*). Coppice oak for edible mushroom cultivation was intensively damaged by a combination of strong absorption of radiocesium to bark and higher transfer factor of radiocesium by mushroom. Rehabilitation of contaminated forest ecosystem for human use was discussed.

Keywords Decontamination · Forest ecosystem · Future perspective · Radiological contamination

---

N. Kaneko (✉)
Faculty of Food and Agricultural Sciences, Fukushima University, Fukushima, Japan
e-mail: kaneko-nobuhirro@agri.fukushima-u.ac.jp

T. Ohkubo
School of Agriculture, Utsunomiya University, Utsunomiya, Tochigi, Japan
e-mail: ohkubo@cc.utsunomiya-u.ac.jp

N. Hijii · C. Takenaka
Graduate School of Bioagricultural Sciences, Nagoya University, Nagoya, Aichi, Japan
e-mail: hijii@agr.nagoya-u.ac.jp; chisato@agr.nagoya-u.ac.jp

C. Takenaka et al. (eds.), *Radiocesium Dynamics in a Japanese Forest Ecosystem*,
https://doi.org/10.1007/978-981-13-8606-0_14

14.1 Introduction

The level of radiological accident of Fukushima Daiichi Nuclear Power Plant (FDNPP) was a "major accident (level 7)" and was the same as the level of Chernobyl Nuclear Power Plant (ChNPP) accident. The radiological contaminants were released from the reactor by venting and hydrogen explosion after the earthquake due the loss of electricity. What is the difference between the two accidents from the viewpoint of environmental contamination by radiological nuclei? As shown in Chap. 1, the major contaminants in Fukushima were radiocesium, whereas there were other contaminants involved in Chernobyl where the reactor was destroyed and nuclear fuels were released from the reactor to the environment. This meant that the vast studies on dynamics and behavior of radiocesium in the environment have accumulated since the major environmental contamination by radiocesium.

Detailed studies were conducted, especially soon after FDNPP accident and until 5 years, when rather rapid change in dynamics of radio-cesium in the environment. Also, forests were one of the major focuses of study, and there were many ecosystem-level studies. Some of these studies were published in the special issue of *Journal of Forest Research* (Ohkubo et al. 2018).

What is the effect of radiocesium contamination to forest organisms? Some observations of deformity of a gall-forming aphid (Akimoto 2014) and a lycaenid butterfly (Hiyama et al. 2012) were published; however, the rate of deformity was declined in 2013 compared to 2012 (Akimoto 2014). Assessment of frogs 18 months after FDNPP accident detected any genetical effects by radiological contamination (Matsushima et al. 2015). Bird population changes were observed in places under 0.3–97 μGy/h (Garnier-Laplace et al. 2015), and the higher exposer overlapped the range of reproduction failure by radiation (42–420 μGy/h). These contamination levels were not observed outside of the evacuation area. We set our study sites outside of the evacuation area; therefore, we presumed that there was no obvious effect of radiological contamination to the population and community of forest organisms. This means that the radiocesium dynamics in forest ecosystem is organized by ecological interaction based on physicochemical cycling of elements in forest ecosystem.

Forest covers 68.5% in area of Japan, and it is an important habitat of various organisms. Forest ecosystem is a complex system and population, and ecosystem approach to risk assessment of environmental radiological contamination needs multidisciplinary cooperation (Bréchignac et al. 2016). Thus, we focused on the comparison of ChNPP and FDNPP accidents (Chap. 1), distribution and redistribution of radiocesium in forest after the fallout (Chap. 2), and contamination levels of trees and wild plants (Chap. 3). Special concerns were movement of radiocesium from bark to tree body of dominant tree species (Japanese cedar, Konara oak, and Japanese red pine), because of their dominance in the forest (Chaps. 4 and 5). Japanese cedar (Sugi) is endemic to Japan and a major plantation tree in mainland Japan. Therefore, its biological trait is affected by the movement of radiocesium.

Moso bamboo was introduced from China, and same as Sugi, it was not found in Europe, so their contamination gives novel information (Chap. 6). Genetic mechanism of transfer of radiocesium in plant body might be applied to understand the behavior of radiocesium in tree plants (Chap. 7). Beside the plants, soil animals (Chap. 10) may link soil contamination to aboveground food web via predatory spiders (Chap. 11), and vertebrate contamination was observed in forest. Natural resource use in the forest was strongly affected by the accident. Tree litter use as material of agricultural compost was evaluated in the forest (Chap. 8) and in laboratory incubation (Chap. 9). Finally, we tried to simulate the fate of contaminants in forest ecosystem using mathematical modeling (Chap. 13).

In this project, we could not cover genetic effects on forest organisms and contamination to freshwater fishes and mushrooms. Contaminated litter has been shown to transfer radiocesium to river food web (Teramage et al. 2014). Coppice oak log is a low material to grow mushrooms. Production of logs from Fukushima Prefecture was long time top in Japan; however, the use of logs has been stopped since the accident. We could not cover this topic.

14.2 Initial Process After the Accident

The initial phase of contamination of forest in Fukushima was very similar to the process observed in Chernobyl. The contamination was not proportional to the distance from the damaged reactors; instead, the several plumes emitted from the reactors contaminated narrow but long distant band of area. The plume consisted different ratios of dry and wet depositions, and this difference affected the fate of radiocesium in the forest (Chap. 2). Kanasashi et al. (2016) found the difference in mobility of radiocesium in Sugi leaves and discussed the effect of chemical form of radiocesium. "Cesium ball (spherical cesium-bearing particles)" is not soluble and was probably contained in the dry deposition (Adachi et al. 2013). It will stay stable in the forest, and future chemical weathering might increase absorption by plants; however, time course change is not estimated.

The accident happened in the middle of March in temperate forest area. Therefore, deciduous broadleaf trees were leafless, so that both the canopy and stem were contaminated, whereas only the foliage ofcanopy of evergreen trees (mostly conifers) were contaminated by the fall-out. Quite a characteristic feature of the forest around FDNPP is that Sugi plantation was dominant in the area. Sugi absorbed radiocesium from the bark and easily transferred it to the heartwood (Chaps. 4 and 5, (Wang et al. 2016; Iizuka et al. 2018; Ogawa et al. 2016). If the radiocesium stayed in the bark and does not move to the wood, the contaminant will move to the forest floor as litter after some years; however, direct absorption increases internal dose of radiocesium. Therefore, the use of timber of Sugi trees needs attention at the time of cutting. Another specific biology of Sugi is that the needle leaves stay on the twig even after leaf decease and the twig stays several years after leaf decease. Therefore it might take a longer time for the radiocesium contamination in the Sugi canopy to move compared to pine (Yoschenko et al. 2017). Several studies have been conducted for Japanese red pine and Konara oak (Ohashi et al. 2014).

There was rather a rapid movement of aboveground radiocesium to forest floor by rain and litterfall. Therefore, both plant contamination by root uptake from soil and food web contamination based on soil organisms will be progressively important with time. Biomagnification of radiocesium is known to be uncommon among organisms, but the exception is mushroom (Calmon et al. 2009; Nakai et al. 2014). This is also confirmed in one species of saprotrophic fungus growing on a fallen twig of Sugi (Chap. 10). In contrast, Japanese deer showed litter increase of radiocesium (Chap. 12). This means the deer is disconnected from detrital food web.

14.3 Changes in Ecosystem Services

We presumed that there was no obvious change in abundance, species composition, and biomass of forest organisms in the forests less contaminated. Therefore, the ecosystem services supported by forest organisms were not affected by the accident (but see Garnier-Laplace et al. 2015). In contrast, change in human activities in forests, such as timber cutting and restriction to the forests, will affect various aspects of forest ecosystem, for example, changes in age structure of canopy trees by reduced cutting activity of forests and changes in food resource and habitat rages in wild mammals by reduced human activities in forests.

The monitoring of radiocesium concentrations in wild plants and mushrooms has been conducted by the local government. There are increasing trends of reduction of contamination. The monitoring effort imposes a burden on locals, but the monitoring activity improves utilization of forest ecosystem services once restricted its access.

14.4 Obstacles to Forest Utilization

The radiocesium cycling in forest ecosystem in Fukushima is approaching quasi-equilibrium after years (Yoschenko et al. 2018). Therefore, ratio between radio- and stable cesium will be an indicator of the fate of radiocesium in forest ecosystem (Yoshida et al. 2004). Slow but steady contamination of forest imposes an obstacle to forest utilization. Concerns are, for example, expected exposure of radiation to forest workers and utilization of forest products such as wild plants and mushrooms for food and timber for materials.

Heavily contaminated croplands have been decontaminated by replacing surface soil with clean soil and deep vertical cultivation. In contrast to the cropland, except the area adjacent to houses (ca. 20 m), no decontamination activities have been done in forest. Standard exposer for fieldworkers is set to 2.5 μSv/h, and it is recommended to use vehicles to reduce the exposer of fieldworkers.

After 1 year of FDNPP accident, the Ministry of Health, Labour and Welfare set the food standard level of radiocesium contamination to 100 Bq/kg (except water). This regulation is much lower than the standards set in EU and the USA. After the decontamination activities in agricultural field, where decontamination and potassium application are conducted to reduce radiocesium absorption, the contamination level of agricultural products was less than 100 Bq/kg. For example, a total of 9,015,099 bags of brown rice (30 kg per bag) produced in Fukushima Prefecture were measured immediately after harvest in 2018, and only 8 bags showed 25–50 Bq/kg, and the others were less than 25 Bq/kg. This means that despite the widespread contamination by radiocesium to croplands outside the evacuation area, the control to suppress the transfer of radiocesium to crops was successful.

The same food standard of radiocesium contamination is applied to wild edible plants, mushrooms, and wild animals from forest. Our study did not cover most of these products; however, the sprout of *C. sciadophylloides* is a popular spring delicacy showing very high transfer ratio of radiocesium (Sugiura et al. 2016), and the role of symbiotic microorganism is estimated to effectively act to transfer radiocesium to the plants (Sugiura et al. 2016; Yamaji et al. 2016).

Widespread but heterogeneous aerial contamination by radiocesium emitted from FDNPP will not change its horizontal distribution; instead, its radiological activity declines according to the physical degradation. The canopy contamination declined rapidly and moved to forest floor, and then most of the contaminants in the forest ecosystem rapidly accumulated in the very surface mineral layer of soil. If the absorption of radiocesium from soil is not substantial, the aboveground part of forest plants will show very low concentration of radiocesium.

There is a voluntary activity of measuring surface gamma-ray strength of timber in Fukushima Prefecture, and it has been under 100 c.p.m. Also voluntary regulation of contamination level allowing cutting tree is set to 05 μSv/h of aerial radiation. This is based on the assumption that under this aerial level, Sugi bark contamination will be lower than 8000 Bq/kg, which is the regulation value of waste radiological contaminated. The absorption of radiocesium from soil and translocation to heartwood and bark litterfall after years may change internal distribution of radiocesium in Sugi tress. Therefore, continuous re-evaluation of standard is necessary for Sugi, the most important timber species in Japan with unique biology.

The standard of radiocesium contamination for Konara oak log for mushroom growing is 50 Bq/kg. This is because of the biomagnification of shiitake mushroom cultivated using Konara oak log. Coppice management is suitable for growing the logs for mushroom cultivation, and the management of coppice in Fukushima Prefecture has been supplied with major part of log use. The coppice rotation was rather short (ca. 15–20 years); however, the management has been terminated after FDNPP accident. Delay of coppice regeneration leads to higher mortality of coppice after cutting. A novel management and use of these broad-leaved trees is needed.

Decontamination of forest has been tested by thinning trees and scraping surface soil with litter on the forest floor. IAEA recommendation was to avoid any management in forest area because of huge economic cost and less human exposure risk. Putting organic matter on the forest floor fosters fungal growth, and fungi translocate radiocesium from soil to the upper organic layer (Fukuyama and Takenaka 2004; Huang et al. 2015). Mycoremediation has been proposed as a promising method to decontaminate forest soil. Phytoremediation using *C. sciadophylloides* is another option to be tested. Biological decontamination, slow but cost-efficient, should be studied for forest rehabilitation.

14.5 Lessons of FDNPP Accident

We should avoid further accidents like ChNPP and FDNPP. Therefore we need to learn a lesson from FDNPP accident, which contaminated the ecosystems.

The difference in species composition of forest canopy trees influences the distribution of pollutants after fallout to forest, because the shape and phenology of canopy depend on tree species and physiognomy and then those affect dynamics of fallout. If a fallout reaches in summer, leaf contamination of deciduous tree species affects the food web relying in those species.

Forest is just like air filter, and its efficiency of trapping pollutants will be higher than other vegetation types. Although, the trapping efficiency of canopy may depend on age and standing crop of forests, because the mineral soil layer strongly absorbs radiocesium, forest acts as temporal reservoir of radiocesium in the terrestrial ecosystem. Cutting windbreak trees around houses reduced aerial dose. Therefore, immediate cutting of trees may reduce further contamination of soil. Burning contaminated biomass can reduce volume of materials. Trees can be preserved after cutting under dry condition. Storing cut trees and further burning can reduce aerial dose for human activities and can control radiocesium.

The major landscape contaminated by the FDNPP accident was mostly forest, but the valleys in the forest were mostly rice paddies. This land use can be found in many Asian countries where monsoon climate is dominant. Forest supports various ecosystem services; it is not only a water source, but also it has been a source of nutrient and fuels for paddy and rural life. Thus, there is a tight nutrient cycling between the forest and paddy, and it seems to be a highly sustainable natural resource use. Artificial radionuclides invaded the nutrient cycling in forest without harming ecological interactions between component organisms. A long-term monitoring of not only the aerial dose but also contamination level of forest components is needed in the contaminated forests.

References

Adachi K, Kajino M, Zaizen Y, Igarashi Y (2013) Emission of spherical cesium-bearing particles from an early stage of the Fukushima nuclear accident. Sci Rep 15:12–15. https://doi.org/10.1038/srep02554

Akimoto S (2014) Morphological abnormalities in gall-forming aphids in a radiation-contaminated area near Fukushima Daiichi : selective impact of fallout ? Ecol Evol 4:355–369. https://doi.org/10.1002/ece3.949

Bréchignac F, Oughton D, Mays C et al (2016) Addressing ecological effects of radiation on populations and ecosystems to improve protection of the environment against radiation: agreed statements from a consensus symposium. J Environ Radioact 158–159:21–29. https://doi.org/10.1016/j.jenvrad.2016.03.021

Calmon P, Thiry Y, Zibold G et al (2009) Transfer parameter values in temperate forest ecosystems: a review. J Environ Radioact 100:757–766. https://doi.org/10.1016/j.jenvrad.2008.11.005

Fukuyama T, Takenaka C (2004) Upward mobilization of 137Cs in surface soils of Chamaecyparis obtusa Sieb. et Zucc. (hinoki) plantation in Japan. Sci Total Environ 318:187–195. https://doi.org/10.1016/s0048-9697(03)00366-8

Garnier-Laplace J, Beaugelin-Seiller K, Della-Vedova C et al (2015) Radiological dose reconstruction for birds reconciles outcomes of Fukushima with knowledge of dose-effect relationships. Sci Rep 5:16594. https://doi.org/10.1038/srep16594

Hiyama A, Nohara C, Kinjo S, et al (2012) The biological impacts of the Fukushima nuclear accident on the pale grass blue. https://doi.org/10.1038/srep00570

Huang Y, Kaneko N, Nakamori T et al (2015) Radiocesium immobilization to leaf litter by fungi during first-year decomposition in a deciduous forest in Fukushima. J Environ Radioact 152:28–34. https://doi.org/10.1016/j.jenvrad.2015.11.002

Iizuka K, Toya N, Ohshima J, Ishiguri F (2018) Relationship between 137 Cs concentration and potassium content in stem wood of Japanese cedar (Cryptomeria japonica). J Wood Sci 64:59–64. https://doi.org/10.1007/s10086-017-1673-9

Kanasashi T, Takenaka C, Sugiura Y (2016) Inferring the chemical form of 137 Cs deposited by the Fukushima Dai-ichi nuclear power plant accident by measuring 137 Cs incorporated into needle leaves and male cones of Japanese cedar trees. Sci Total Environ 553:643–649. https://doi.org/10.1016/j.scitotenv.2016.02.118

Matsushima N, Ihara S, Takase M, Horiguchi T (2015) Assessment of radiocesium contamination in frogs 18 months after the Fukushima Daiichi nuclear disaster. Sci Rep 5:9712. https://doi.org/10.1038/srep09712

Nakai W, Okada N, Ohashi S, Tanaka A (2014) Evaluation of 137Cs accumulation by mushrooms and trees based on the aggregated transfer factor. J Radioanal Nucl Chem 303:2379–2389. https://doi.org/10.1007/s10967-014-3729-2

Ogawa H, Hirano Y, Igei S et al (2016) Changes in the distribution of radiocesium in the wood of Japanese cedar trees from 2011 to 2013. J Environ Radioact 161:51–57. https://doi.org/10.1016/j.jenvrad.2015.12.021

Ohashi S, Okada N, Tanaka A et al (2014) Radial and vertical distributions of radiocesium in tree stems of Pinus densiflora and Quercus serrata 1.5 y after the Fukushima nuclear disaster. J Environ Radioact 134:54–60. https://doi.org/10.1016/j.jenvrad.2014.03.001

Ohkubo T, Kaneko N, Kaneko S et al (2018) Radiocesium dynamics in forest ecosystems after the Fukushima Nuclear Power Plant accident: experiences during the initial five years. J For Res 23:1–2. https://doi.org/10.1080/13416979.2018.1429733

Sugiura Y, Kanasashi T, Ogata Y et al (2016) Radiocesium accumulation properties of Chengiopanax sciadophylloides. J Environ Radioact 151(Pt 1):250–257. https://doi.org/10.1016/j.jenvrad.2015.10.021

Teramage MT, Onda Y, Kato H, Gomi T (2014) The role of litterfall in transferring Fukushima-derived radiocesium to a coniferous forest floor. Sci Total Environ 490:435–439. https://doi.org/10.1016/j.scitotenv.2014.05.034

Wang W, Hanai Y, Takenaka C et al (2016) Cesium absorption through bark of Japanese cedar (Cryptomeria japonica). AU – Wang, Wei J For Res 21:251–258. https://doi.org/10.1007/s10310-016-0534-5

Yamaji K, Nagata S, Haruma T et al (2016) Root endophytic bacteria of a 137Cs and Mn accumulator plant, Eleutherococcus sciadophylloides, increase 137Cs and Mn desorption in the soil. J Environ Radioact 153:112–119. https://doi.org/10.1016/j.jenvrad.2015.12.015

Yoschenko V, Takase T, Hinton TG et al (2018) Radioactive and stable cesium isotope distributions and dynamics in Japanese cedar forests. J Environ Radioact 186:34–44. https://doi.org/10.1016/j.jenvrad.2017.09.026

Yoschenko V, Takase T, Konoplev A et al (2017) Radiocesium distribution and fluxes in the typical Cryptomeria japonica forest at the late stage after the accident at Fukushima Dai-Ichi Nuclear Power Plant. J Environ Radioact 166:45–55. https://doi.org/10.1016/j.jenvrad.2016.02.017

Yoshida S, Muramatsu Y, Dvornik AM et al (2004) Equilibrium of radiocesium with stable cesium within the biological cycle of contaminated forest ecosystems. J Environ Radioact 75:301–313. https://doi.org/10.1016/j.jenvrad.2003.12.008

MIX
Papier aus verantwortungsvollen Quellen
Paper from responsible sources
FSC® C105338

If you have any concerns about our products,
you can contact us on
ProductSafety@springernature.com

In case Publisher is established outside the EU,
the EU authorized representative is:
Springer Nature Customer Service Center GmbH
Europaplatz 3, 69115 Heidelberg, Germany

Printed by Libri Plureos GmbH
in Hamburg, Germany